The Land That Could Be

Urban and Industrial Environments
Series Editor: Professor Robert Gottlieb, School of Public Policy and Social Research, University of California, Los Angeles

The U.S. Paper Industry and Sustainable Production: An Argument for Restructuring, Maureen Smith, 1997

Human Settlements and Planning for Ecological Sustainability: The Case of Mexico City, Keith Pezzoli, 1998

Greening the Ivory Tower: Improving the Environmental Track Record of Universities, Sarah Hammond Creighton, 1998

Making Microchips: Policy, Globalization, and Economic Restructuring in the Semiconductor Industry, Jan Mazurek, 1998

The Land That Could Be: Environmentalism and Democracy in the Twenty-First Century, William A. Shutkin, 2000

The Land That Could Be
Environmentalism and Democracy in the Twenty-First Century

William A. Shutkin

The MIT Press
Cambridge, Massachusetts
London, England

First MIT Press paperback edition, 2001

This book was set in Sabon by The MIT Press and printed and bound in the United States of America.

Printed on recycled paper.

The writing of this book was supported by:
The Florence and John Schumann Foundation
Furthermore, the publication program of the J. M. Kaplan Fund
The Nathan Cummings Foundation

Library of Congress Cataloging-in-Publication Data

Shutkin, William A.
The land that could be : environmentalism and democracy in the twenty-first century / William A. Shutkin.
p. cm. -- (Urban and industrial environments)
Includes bibliographical references and index.
ISBN 0-262-19435-X (hc : alk. paper), 0-262-69270-8 (pb)
1. Environmentalism--United States. 2. Democracy--United States. I. Title. II. Series.
GE197.S48 2000
363.7'00973--dc21 99-056829
CIP

To my parents, Elizabeth and Peter Shutkin, and my wife and daughter, Sally Handy and Olivia Shutkin, whose collective devotion are the rich, loamy soil that sustains me and whose legacy is the reason for this book

Contents

Foreword by David Brower ix
Preface xiii
Acknowledgments xix

Introduction 1

1 All Things Merge into One: Environmentalism and Civic Life 21

2 The Physics of Civics, or The Environmental Consequences of Civic Decline 45

3 The Land That Could Be: American Environmentalism and the Pursuit of Sustainable Communities 89

4 Urban Agriculture in Boston's Dudley Neighborhood: A Modern Twist on Jefferson's Dream 143

5 Oakland's Fruitvale Transit Village: Building an Environmentally Sound Vehicle for Neighborhood Revitalization 167

6 Community-Based Conservation and Conservation-Based Development in Rural Colorado 189

7 Smart Growth, Community Planning, and Cooperation in Suburban New Jersey 209

8 Coming Full Circle: An Emerging Model of Environmentalism and Democracy 237

Notes 245
Index 263

Foreword

We can no longer afford the luxury of pessimism. Despair is a sin, as Lord Snow said. Hope is more fun. So is rethinking. We can rethink Progress, Sustainability, Mobility, Design, Conservation, Preservation, Restoration. And we must. For too long, environmentalists have hung their heads in disgust, decrying under their breath the plunders of capitalism and greed; and for too long, those same forces have turned the American landscape, its magnificent canyons and rugged coastlines, its verdant forests and boundless plains, its central cities and rural villages, into a playground for destruction and relentless development. No consideration for consequences or future generations.

We have yet to experience real democracy or real capitalism. If we had, we wouldn't see blighted inner-city neighborhoods where contaminated land and childhood lead poisoning are commonplace, or sprawling suburbs, chopped up into subdivisions and shopping malls; we wouldn't waste our most precious form of capital, our natural resources. The economist Herman Daly has said that under our present economic system, the earth is a business in liquidation. Real democracy and real capitalism value people and nature, do everything possible to protect and preserve communities and ecosystems, and consider the fate of future generations.

We need a short word, of few syllables, to describe the environmental movement. How about three little verbs—care, save, renew? Keep these verbs alive for our kids to celebrate.

Add democracy and capitalism, admired by many, practiced by few. Democracy is now run by the haves. Capitalism betrays its source, nature's services. Add respect for the rights of nature, including ourselves. For example, don't dig plutonium, dump it, or breathe it, or spoil the air that we need to inhale every two minutes or so or die. Remind Detroit and Big Oil.

Capitalists, remember this about nature's services: They are not free of charge. Every year we use three dollars worth of resources for every two dollars we recover in the global equivalent of the GNP. That's a heavy loss of real capital. If we keep ignoring the source, nature may deny us credit.

In the face of this reality, we need to be hopeful and consider alternate strategies to take charge of our fate and live up to our best political and moral commitments in enacting a vision of a better world.

This is what Bill Shutkin's book is all about. It is about taking control of our shared destiny and reclaiming what Langston Hughes called "these great green states." For nearly nine decades, I have seen the power and glory of what we now call environmentalism; I have seen great victories and equally great defeats. But I know that our finest moments, whether as environmentalists or not, lie just ahead of us. If we act.

The Land That Could Be profiles the kind of action that our democracy and environment need most. Shutkin writes not merely as an observer of this action, but someone who himself has been a part of it. He's a dynamic, young environmental leader (a mere half-century younger than myself) who has cut his teeth in the trenches of some of the country's most physically and socially distressed places. He's an environmental visionary looking for and creating solutions to today's problems with a passion that would make John Muir and Martin Luther King equally proud.

The Land That Could Be describes a better way for environmentalists to deal with the challenges posed by our economy and way of life. As the book describes, across the country efforts are underway that herald a new era of environmental activism and democratic renewal, what Bill Shutkin calls "civic environmentalism." From urban, suburban, and rural areas, these efforts are at once about building democratic communities and healthy economies and about protecting the physical places and resources that sustain them.

These stories reflect a pivotal realignment of environmentalism away from exclusively legal or regulatory approaches, based on individual pollutants and single media (air, water, soil) and handed down from on-high, to a more bottom-up approach aimed at overall community and ecosystem health, including social and economic health. Civic environmentalism is more planning-based and preventive than traditional environmental protection strategies, which have tended to focus on the biggest, most readily controllable pollution sources while ignoring the more intractable problems resulting from land use and development activities. It blends legal measures with fiscal policy, good science, governance mechanisms, and plain-old civic will to effect durable, indeed sustainable, environmental and social outcomes wedded to a particular community and a particular place. Most important, civic environmentalism promotes civic engagement, to involve the individuals and businesses who contribute so mightily to our pollution problems yet also hold the key to the solutions.

In the face of the sometimes pernicious forces of global capitalism that continue to externalize environmental and social costs and undermine place-based commitments, a system of private property rights that allows individuals to do with their land largely as they please, and a troubling lack of solid public discourse, civic environmental strategies are taking root across the borders of geography, income, and race and are demonstrating that our environment and community respond to each other, are one.

Models of sustainable development like those described in *The Land That Could Be* are transforming the way we think about our economy and our community. The players in these projects, including environmentalists, are treating the economy and community not as abstract concepts but as rooted in a place, their health inextricably tied to the health of the environment. What the examples in this book show is that environmental values such as pollution prevention, organic agriculture, ecosystem health, and mass transit can actually drive and sustain the economy and social health, providing environmentalists a powerful opportunity to influence the fundamental ways people live and conceive their lives.

The challenge as the twenty-first century dawns is to create a sustainable society and, in the process, restore ecosystems. Only then will we have turned the corner toward a greener century. This book helps lead the way.

David R. Brower, Berkeley, California
Chair, Earth Island Institute; Director, Sierra Club; Founder, Friends of the Earth

Preface

We have given up the understanding . . . that we and our country create one another, depend on one another, are literally part of one another; that our land passes in and out of our bodies just as our bodies pass in and out of our land; that as we and our land are part of one another, so all who are living as neighbors have, human and plant and animal, are part of one another, and so cannot possibly flourish alone; that, therefore, our culture must be our response to our place, our culture and our place are images of each other and inseparable from each other, and so neither can be better than the other.

—Wendell Berry, *The Unsettling of America*

Without the oxygenating breath of the forest, without the clutch of gravity and the tumbled magic of river rapids, we have no distance from technologies, no way of assessing their limitations, no way to keep ourselves from turning into them. . . . Only in regular contact with the tangible ground and sky can we learn how to orient and navigate in the multiple dimensions that now claim us.

—David Abram, *The Spell of the Sensuous*

The dawning of a new century provides a unique opportunity for Americans to take stock. It is the perfect occasion for reckoning and rejuvenation, making serious pledges and promises for the new age, measuring progress toward goals, both personal and public, and collectively asserting a broad vision in the face of withered or unfulfilled social aspirations. This book is offered up in this millennial spirit, at once an accounting of and prophecy about American environmentalism and the promise of democratic communities.

Perhaps it is no accident that just as the twenty-first century begins, environmentalism in the United States finds itself at a watershed, a crossroads. After all, the new millennium is as much a cause as it is a symbol of changing attitudes and practices, not only concerning environmentalism but across the spectrum of social issues. For the first time, environmentalists are addressing issues formerly dismissed or ignored, like the urban environment and working landscapes, and a diverse group of stakeholders—from inner-city activists to suburban municipal officials to rural ranchers—are engaging in environmental protection efforts.

Environmentalism is evolving from an essentially elitist movement accompanied by a complex system of laws and policies fixated on preserving undeveloped land and resources and controlling pollution from major sources to a more democratic call for healthy, sustainable communities across geographic, economic, and cultural lines. Instead of merely reacting to environmental changes and decrying the pollution and waste generated by our liberal capitalist economy, an expanding environmental constituency is devising alternatives to traditional approaches to economic development and environmental protection. One need only look at the host of environmental initiatives launched in the 1990s, and the emergence of a new field of design and engineering called industrial ecology to see the transformation. In their best light, these and other initiatives are aimed at not only cleaning up the nation's air, water, and soil, but making American communities more livable and the economy more sustainable.

I call this new approach *civic environmentalism* because it marries a concern for the physical health of communities with an understanding that part and parcel of environmental quality is overall civic health. Civic environmentalism holds that the physical environment—the places people inhabit or otherwise affect through their actions—influences our behavior and our sense of place, community, and well-being. The environment does much more than simply provide for our material needs, thus sustaining human life. As the urban historian Dolores Hayden explains, all landscapes, and especially urban landscapes, "are storehouses for . . . social memories, because natural features such as hills or harbors, as well as streets, buildings and patterns of settlements, frame the lives of many people and often outlast many lifetimes."[1]

Moreover, our environment, whether in urban centers or remote wilderness areas, is not merely perceived by our five senses, the "shaping perception," as the historian Simon Schama suggests, "that makes the difference between raw matter and landscape,"[2] but is constructed by society to meet its dual needs of economic production (reflected in barns, piers, highways, and factories) and social reproduction (expressed in housing, churches, and schools).[3]

In short, the environment is the sum of all those places in cities, suburbs, and rural areas that play an essential part in constituting our sense of ourselves as individuals and members of a community and that demand our care and attention if they are to enhance, rather than diminish, that sense. To ensure the production and protection of a healthy environment requires the participation of those whose quality of life ultimately depends on it: ordinary citizens.

Such participation need not be heroic. Rather, as Daniel Kemmis, director of the Institute for the Rocky Mountain West, explains, it is the "simple, homely practices which are the last best hope for revival of genuine public life" and, I would add, for the protection of the environment.[4] Similarly, the late Texas congresswoman Barbara Jordan argued, "Democracy cannot be saved by supermen, but only by the unswerving devotion of millions of middle men."[5] The same powerful logic, this book suggests, applies to the environment.

A place worth living in and leaving to future generations requires the investment of time and money, as well as an understanding of the fundamental relation between the quality of the physical environment and society's overall quality of life. Getting there often requires solutions that go beyond the traditional legal or adversarial approach. It requires a shared sense of commitment to a particular place and to the life of a particular community arrived at by rigorous dialogue and the practice of genuine citizenship.

This book asserts that environmentalism is as much about protecting ordinary places as it is about preserving wilderness areas; as much about promoting civic engagement as it is about pursuing environmental litigation; and as much about implementing sound economic development strategies as it is about negotiating global climate change treaties. Ultimately, I believe, environmentalism is nothing less than about our conception of ourselves as a social and political community—what the bald eagle, our national symbol, really means.

The title of this book is meant to suggest the inextricable bond between nature and nation. The "land" referred to in it denotes at once a physical, objective thing comprising plants, animals (human and nonhuman), rivers and mountains, and the abstract ideal of a state, of a people, wedded to a set of noble social and political principles like freedom, equality, and justice. A major premise of this book is that nature and culture—the ideas and practices by which human societies are constructed and maintained—are not mutually exclusive and that, in America at least, the two are fundamentally linked, through ideology and history, if not through the arts, in a uniquely powerful and promising way. It is no accident that the United States was the first country to establish national parks, or that the celebrated and revered public figure, Thomas Jefferson, based his political ideology on the notion of land stewardship. America's history and identity are shot through with environmental themes; our hopes and aspirations are bound up with environmental symbols.

The book's title derives from Langston Hughes's poem, "Let America Be America Again." That poem, parts of which are woven throughout *The Land That Could Be,* captures the fullness and spirit of the relation between the environment and the ideal of American democracy. Like no other American voice before or after him, Hughes sings of the indelibly intertwined fate of the land and its people. The achievement of democratic communities of free and equal citizens and environmental quality are, on Hughes's account, the same project. For Hughes, the land is the preeminent metaphor of freedom, hope, and redemption; it is also literally the place where the American dream resides.

Thus, the book's title speaks to the essential link between the idea of democracy and environmental protection. As linguists George Lakoff and Mark Johnson explain, "The essence of metaphor is understanding and experiencing one kind of thing in terms of another."[6] Metaphors, they argue, structure our concepts, activities, and language. This book interprets the idea of community and solidarity, of civic life itself, by way of the environment and asserts the primacy of the physical places we inhabit to the ways in which we live as democratic citizens. The metaphor of the land points to the fundamental bond between civic engagement and environmental health. To the extent that we live by this metaphor, we achieve not only a strong democracy, to borrow political scientist Benjamin Barber's phrase, but one worth living in, where access to a quality environment is implicit in our democratic ideas and practices, in our praxis.[7]

Thus, this book is about the unique power of environmentalism as a tool to realize the as-yet-unfulfilled potential of this land—in Hughes's words, "the land that never has been yet/and yet must be"—to build from the ground up democratic communities, communities whose physical conditions may finally come to match the quality of the social ideals on which they are based.

Acknowledgments

For most of my adult life I have had the good fortune to be able to do for a living what I love: to celebrate, defend, and restore the natural environment, including human communities, in all their forms and places, and—heeding Ralph Waldo Emerson's injunction that ideas must be published—to write about it, in scholarly articles, op-eds, poems, and now a book.

In any writing project, the author is but a proxy for the scores of people who, at any given time, provided invaluable lessons, support, and criticism. This book is no exception. I am grateful to all those teachers, mentors, and friends who have inspired or indulged my twin passions for the environment and the ideal of American democracy. Many of these people I know only through their writing. Of those whom I have had the privilege to know personally, some are gone. Their words and deeds are reflected in these pages.

In a far more practical sense, I owe a huge debt of gratitude to the people and institutions that made this book possible. First, I thank John and Bill Moyers of the Florence and John Schumann Foundation, Joan Davidson of Furthermore, the publication program of the J. M. Kaplan Fund, and Mark Walters and Charlie Halpern of the Nathan Cummings Foundation for their invaluable financial support. Many authors who choose to write serious, if not topical, books and go with independent or university presses are not afforded this luxury, though they should be. I hope *The Land That Could Be* does their support justice.

Many thanks also go to several MIT Press staff, including Larry Cohen, Clay Morgan, and Enza Vescera, and series editor Bob Gottlieb, each of whom was exceptionally helpful and responsive. In the course of writing the book, I spoke to dozens of individuals who provided information, insight, inspiration, or guidance, or some combination thereof. My heartfelt thanks go to the following: Lois Adams, Mitch Bernard, Lee Breckinridge, Hooper Brooks, David Brower, Mark Burget, Denise Coyle, Thomas Czerniecki, Mikhail Davis, John DeVillars, Mark Dowie, Veronica Eady, Jay Fetcher, Doug Foy, Chris Hudson, Anne Kelly, Barbara Lawrence, Ed Lloyd, Penn Loh, Mindy Lubber, Vernice Miller, C. J. Mucklow, Jeff Olson, Susan Dorsey Otis, Deval Patrick, Zyg Plater, Michael Rios, Carol Rufener, Rusty Russell, Trish Settles, Lynne Sherrod, Mike Tetreault, Greg Watson, Tony Wood, and Marty Zeller.

Special thanks also go to my research assistants, Rafael Mares, a student at Harvard Law School, and Steve McShea, a Boston College Law School student. Their contributions to the case studies were critical. As well, friends Matt and Amy Brand, Kent Greenfield, John Fox, John Bonifaz, and Robin Kelsey were especially generous with their support and insights.

The moral support of parents and friends was, as always, invaluable, but it was my wife, Sally, and our daughter, Olivia, along with our cats, Orbit and Rocket, who really saw me through.

The Land That Could Be

Introduction

In reclaiming and reoccupying lands laid waste by human improvidence or malice . . . the task is to become a co-worker with nature in the reconstruction of the damaged fabric.

—George Perkins Marsh, *Man and Nature*

How shall we turn the ghetto into a vast school? How shall we make every street corner a forum . . . every houseworker and every laborer a demonstrator, a voter, a canvasser and a student? The dignity their jobs may deny them is waiting for them in political and social action.

—Martin Luther King, Jr., "Where Do We Go from Here?"

America was founded on the belief that the health of democracy is inextricably bound up with the bounty and extent of the nation's physical environment. Civic virtue and democratic habits of the heart were believed to be born of the living arrangements found on a small farm or in an agrarian village. Many believed that American communities, to be at their democratic best, need to be arranged in such a way as to connect individuals to each other and to the land, their earthly place. It was the lack of massive urban centers like those found in the Old World of Europe, where anomie, inequality, and decadence went hand-in-hand, that for America's founders enabled American democracy to thrive.[1]

Yet as the twenty-first century begins, there is widespread uncertainty about the social and environmental fate of the American experiment in democracy and about the relevance of inherited belief systems and philosophies like agrarianism in the light of the social and environmental challenges we now face. Mindful of the need not to be nostalgic or sentimental about the past, it is safe to say that contemporary American society, for all it has gained in terms of economic prosperity, technological prowess, and social justice, has lost a lot too.[2]

Leading cultural critics and social scientists have observed a growing sense among ordinary Americans that they have lost control over their own lives, that the economic, political, and cultural disruptions of modern life have eroded Americans' civic sensibility and willingness to participate in public affairs.[3] The impersonal power of the marketplace and its megacorporations, the rapid rise of ever-faster digital technologies, and the globalization of communication and commerce have had a destabilizing effect on many Americans, who are struggling to keep abreast of the enormous changes these forces have wrought.

Meanwhile, the traditional role of government, at least since the New Deal, as the overseer of the general welfare and the provider of last resort has been significantly called into question, if not fundamentally altered. In the current period of devolution and decentralization of government authority spawned during the Republican ascendancy to power under Ronald Reagan in the 1980s and later the Republican-dominated Congress in the 1990s, most Americans are still trying to figure out just what the new role of government is or ought to be, and where the private realm ends and the public begins. *Privatization* became the buzzword of the 1990s, as federal, state, and local governments looked to individuals and the private sector for solutions to social problems once exclusively reserved for government, from crime and welfare to education and the environment. Some are left to wonder whether there is any public life at all left in America.[4]

At the same time, despite the favorable economic conditions of the 1990s, the gap between rich and poor, the haves and the have-nots, the knowledge class and the labor class, is greater than ever, and growing at a steady pace. Notwithstanding strong employment rates and increases in earned income, lower- and middle-income Americans have not kept pace with the burgeoning economy, facing stagnating wages and higher costs of living. As a groundbreaking study by the policy think tank Redefining Progress reports, though the gross domestic product might be up, many Americans are feeling down.[5]

In addition, racial inequality and segregation persist, mirroring the striking economic disparity among Americans. Although considerable progress has been made in the last half-century in the age-old struggle to secure the civil rights of minorities, people of color in the United States continue to lag behind their white counterparts in their pursuit of the American dream. From education and income to health and environmental quality, people of color still face formidable deficits in trying to improve their quality of life.

A central argument of this book is that part and parcel of this diminution of civic spirit and rise in economic and social inequality has been the deterioration of the American environment, both built and undeveloped. As even the most casual observer can attest, the environment of many of America's cities, suburbs, and rural areas has been under assault from an array of environmental harms resulting from overdevelopment, automobile emissions, and industrial pollution, among other causes.[6] We live, it seems, in a physical world no longer designed by mother nature, where humans' handiwork is the exception, not the rule, but instead by engineers, industrialists, developers, and planners more concerned about the bottom line than their environmental legacy.

Environmental changes brought about by social and economic forces have had a corrosive effect on our sense of place and self, and have divided communities along racial and economic lines as well as cutting them off from access to safe, unspoiled places.[7] It is no accident that with America's economic growth has come environmental harm and the displacement and segregation of human communities. This is, after all, the longest running of American themes, originating with the European settlement of precolonial New England, where significant numbers of Native Americans and wildlife species were extinguished and forests denuded almost overnight, and continuing through the centuries, most recently in the form of urban renewal in the 1940s and 1950s, and suburban sprawl in the 1980s and 1990s.[8] Despite considerable progress in cleaning up the nation's air, water, and soil over the past three decades, pervasive environmental degradation persists: countless acres of contaminated land, miles upon miles of rivers and streams unsafe for swimming or fishing, toxic haze blanketing entire regions, thousands of native plant species disappearing every year, and millions of children afflicted with lead poisoning and asthma from indoor and outdoor air pollution, among other pollution problems.

We have thus arrived at the beginning of the twenty-first century at a place far afield from eighteenth-century America's pristine agrarian villages. A largely and increasingly urban society (though in significant numbers many today are fleeing the cities in pursuit of idyllic pastoral settings),[9] we seem to have lost our traditional moorings and the accompanying sense of confidence about who we are as a people and where we are headed. Once a proud agrarian republic, then a pioneering industrial democracy, we inhabit today what many call, for want of a better descriptor, the postindustrial order, an unstable alloy of old metropolises and new, of edge cities and third-ring suburbs, of factories, malls, and subdivisions, and of working farms and fields. As always, there remains that awesome space, the American wilderness, an endless source of national mythology and pride amid the wrenching changes in the American landscape of the past half-century.

An Emerging Model of Environmentalism

In response to this apparent physical and social crisis of democracy, I cofounded in 1993 a nonprofit environmental law and education center, Alternatives for Community & Environment (ACE), dedicated to helping residents of lower-income neighborhoods and neighborhoods of color in and around Boston protect and improve their environment as a means of helping to revitalize and reclaim their communities. The moral impulse behind the idea of ACE was developed at an early age. I am a child of the American suburbs of the 1960s and 1970s and of the Jewish upper-middle class. My parents, like many of their peers, fled New York City in the mid-1960s to settle the then-new frontier, at least for ethnic minorities, of southwestern Connecticut. I was raised in Stamford, a historically blue-collar, suburban city located in one of the wealthiest counties in the United States. An immigrant community, Stamford has always had its share of haves and have-nots. It hosts a mix of blue- and white-collar residents, of elite private and struggling public schools, of urban and semirural settings, of whites and people of color, of Jews, Muslims, Catholics, and Christians. I attended both private and public schools, trekked on urban streets and in nature preserves, played Little League baseball with whites, blacks, Latinos, and Asians, sons of plumbers and of lawyers, and attended Catholic mass on more than a few occasions.

Moving between and among these different subcultures and social spaces that in many ways define the postmodern, multicultural age, I became a rather introspective young man, heeding Socrates' injunction that "the unexamined life is not worth living." I began to take seriously the call to service and the moral responsibility my parents, better teachers, and favorite literature described but too few of my role models exemplified. My morality was forged by, on the one hand, my love of the environment, having grown up next door to a 200-acre nature sanctuary and been given by my parents the opportunity to explore many of the country's natural wonders, and, on the other, my experiences in inner-city Stamford, where I spent considerable time after school in my teenage years teaching guitar to young students from one of the city's housing projects. The project abuts Interstate 95, and like most other public housing stock of its era is dangerous and dilapidated, a collection of brick boxes more like a prison facility than a place to call home. The projects were anathema to my moral and aesthetic sensibility. It was not that the rest of the suburban environment was much better, with its sprawl, sanitized corporate headquarters, traffic congestion, and isolated, dead-end streets. But the access I enjoyed to beautiful places in and around the city was missing for these kids, whose brothers and sisters were my high school classmates.

Some twelve years later ACE was born. The organization appropriates and institutionalizes the agrarian notion (with a decidedly urban, multicultural twist) that the physical condition of America's communities is a critical factor in the nation's success as a robust democratic republic. In other words, ACE embodies the idea that healthy, vibrant social and political life presupposes a bare minimum of environmental quality. Community, democratic community, and environmentalism are one.

Because environmental issues tend to defy political, cultural, and geographic borders (consider the pervasive, insidious effects of regional air pollution, suburban sprawl, or contaminated drinking water), they are often a unique way to bring people together from all walks of life and backgrounds. They provide an opportunity to realize the ideal of community: diverse individuals and groups coming together around a common concern and giving voice to a vision for the common good. ACE was designed to capitalize on this opportunity. Further, ACE strives to promote the idea that ordinary people, with enough information and technical support at their disposal, can and must participate in the decision making that affects their lives. At bottom, informed, meaningful citizen participation is what democracy is all about.

In no small way, ACE represents a response to the failure of traditional environmentalism to articulate and act on a democratic social vision. ACE is really nothing more than the kind of community resource most affluent white Americans have enjoyed for decades. Because environmental protection, like all other social issues, is a matter of the distribution of benefits and burdens, those who hold the most power tend to receive the most benefits, such as cleaner air, access to parks, and rigorous enforcement of environmental laws. Those without political or economic power (the two go hand-in-hand) tend to bear the brunt of environmental burdens, like polluting facilities (incinerators, auto body shops, trash transfer stations, and the like), contaminated land and buildings, and underenforcement of environmental laws.

Traditionally, environmental protection has been a zero-sum game in which the success of the wealthy in fighting or avoiding environmental problems has resulted in the displacement or persistence of those problems in lower-income and minority communities. The siting of landfills and incinerators (whose solid waste feedstock often comes from sources outside the host community), abandonment of contaminated land (usually by those who could afford to pick up and leave), and air quality hot spots in urban neighborhoods (caused largely by emissions from pass-through, suburban commuter vehicles) are some of the consequences borne by the less affluent resulting from the actions of the more affluent.

ACE attempts to level the playing field by working with residents and organizations in hard-hit neighborhoods to advocate for environmental protection and public health. Often this means battling against unwanted development or compelling a private actor or government agency to clean up a polluted piece of land or dilapidated building that has been ignored for decades. In a sense, ACE helps hold the line against further deterioration in environmental quality and human health. In the communities where it works, like the hard-hit Boston neighborhoods of Roxbury and Chinatown, environmental degradation and high morbidity and mortality rates for environmentally related diseases such as asthma are all-too-common indicators of community health.

But more than simply holding the line, ACE and its community constituents are attempting to change the way environmental protection is conceived. Rather than calling for pollution burdens to be redistributed to some other communities, ACE advocates for pollution-prevention strategies so that no one has to face the threat of undue environmental harm. Moreover, ACE promotes the idea that environmental decision making must itself be transparent and open to those who for too long have been left out, believing that those best able to protect their environment are those living in it.

As successful as ACE has been in helping to improve the environmental and public health conditions in communities of color in and around Boston, significant challenges remain. For example, for several years beginning in 1993, ACE helped a number of community organizations fight a proposed asphalt plant in Boston's South Bay section. A massive filled tideland, the South Bay is the heart of the city's historic industrial corridor and sits at the crossroads of five diverse neighborhoods: Chinatown, Dorchester, Roxbury, South Boston, and the South End. It abuts the Southeast Expressway, the main north-south roadway feeding Boston's notorious elevated highway Central Artery. On an average day, tens of thousands of South Shore commuters slouching toward work or home contribute to the South Bay's persistent air pollution problems. Public health experts call the South Bay the "zone of death" because of the extraordinary mortality rates for upper respiratory diseases and other health effects associated with poor air quality.

The South Bay area and its adjacent neighborhoods are full of vacant, polluted land (over fifty confirmed hazardous waste sites are located in Roxbury's Dudley neighborhood alone) and businesses, such as waste transfer stations and auto body shops, whose operations add to the area's environmental degradation. For decades, these neighborhoods have suffered high rates of unemployment and poverty. Even today, despite Boston's economic boom, some South Bay neighborhoods are experiencing unemployment rates as high as 20 percent. The area lies within the city's empowerment zone, the federal program intended to attract much-needed jobs and development to economically distressed communities.

The proposed asphalt plant, owned by the Boston-based Todesca Equipment Company, offered only four full-time jobs yet a host of environmental hazards, from truck traffic and emissions to air quality impacts from the facility itself. More important, the 2.7-acre site presents a potentially attractive opportunity for job-generating, "green" redevelopment—the kind of economic development the neighborhoods have been promoting for years.

Known as the Coalition Against the Asphalt Plant (CAAP), the community groups opposed to the plant, from each of the neighborhoods abutting the South Bay, worked with ACE to stop the project. From lawsuits to administrative appeals to direct action, CAAP organized relentlessly, pulling together an unprecedented campaign based on the frustrations and anger of tens of thousands of inner-city residents tired of being steamrolled by unwanted development. CAAP was not opposed to development per se. In fact, its members explicitly called for jobs and business growth. But they would not settle for a polluting facility that offered almost no economic opportunity for local residents.

What finally killed the Todesca proposal, at least until the proponent's lawyers appealed, was a ruling by Boston's Board of Health in the spring of 1996 prohibiting the use because of the area's extraordinary public health conditions. The board took seriously the South Bay's reputation as the "zone of death" and ruled in effect that "enough is enough," heeding the cries of the neighborhood groups. That the Board of Health had to step in is telling. Because of the facility's relatively small emissions levels ("only" approximately 13 tons of air pollution per year, according to the proponent's figures), recourse to environmental laws to stop the project was extremely limited. CAAP had to resort to extraordinary claims based on the area's dangerous environmental and public health conditions to get environmental regulators' attention, and even then it was limited to a few technical issues regarding the plant's engineering specifications for its allegedly state-of-the-art pollution control equipment. Moreover, because environmental laws confer no jurisdiction on environmental officials over local land use decisions, the original action granting the plant a permit to be built on the South Bay site was essentially unreviewable, except by the same zoning board that made it in the first place.

This gap in the system of environmental laws meant that although South Bay residents faced greater exposure to environmental hazards than most of their Boston neighbors due to the cumulative effects of decades of pollution, their claims were essentially meaningless. The Board of Health, however, like most other such boards, has broad authority dating back to the nineteenth century, when the only governmental bodies established to deal with what we now call environmental problems were the local health boards, acting under the state's police power to protect the general welfare. In those days, the problems typically had to do with odors from pig farms, leather tanneries, and other "obnoxious" or "injurious" operations. In an age of sophisticated pollution control technology and a vast complex of environmental regulations, it is thus supremely ironic that but for the nineteenth-century public health laws, the asphalt plant and all similar facilities would face no significant legal hurdles.

Despite the victory, the fate of the site and the asphalt plant still hangs in the balance. CAAP won the battle but stands to lose the war. It is this fact that underscores the ultimate challenge facing CAAP and all other communities that want to take control of their environmental and economic destiny against the tide of historic injustices, social change, and often subversive market forces. Unable to reposition itself from an oppositional force to a catalyst for acceptable economic development, CAAP can only stand on the sidelines while others determine the future of the site. Since the Board of Health decision, several other proposals have been floated, from a large surface parking lot for workers in the nearby Longwood Medical Area, to a tire recycling plant. None has drawn much support. CAAP's campaign, like so many other community-led initiatives, has been essentially powerless to implement its own vision for the South Bay, grounded in a desire to improve economic conditions for area residents while protecting and restoring environmental quality.

Meanwhile, other opportunities to improve neighborhood conditions beckon. Just across the street from the asphalt plant site on South Bay Avenue sits the South Bay incinerator site, a now-vacant 3.3-acre parcel that for over twenty-five years was home to the city's municipal solid waste incinerator. In 1975, the facility was shut down by court order due to the environmental hazards it created. Only in 1997, after twenty-two years of sitting idle like a mammoth brick sarcophagus, was the facility demolished in response to community pressure supported by ACE. Many of the same groups that organized in opposition to the asphalt plant came together to compel the Commonwealth of Massachusetts, the owner of the site, to clean up the incinerator site. At the same time, they successfully advocated against the siting of a proposed "megaplex"—a huge, co-located convention center–stadium facility—for the incinerator and nearby South Bay sites. Calling themselves Neighborhoods United for the South Bay (NUSB), they, like CAAP, want clean, job-generating development in the South Bay. However, the Suffolk County House of Corrections, located next door to the incinerator site, has eyed the parcel as a potential site for a new central lockup facility, made necessary by the county's skyrocketing prison population. A *Boston Globe* editorial supported the prison proposal, claiming that manufacturing and industrial uses were not feasible for the area.

A stone's throw from the South Bay sites in Roxbury is a massive concrete box, home to the abandoned Stride Rite manufacturing facility. Built in the fortress style of the 1970s, the empty facility is all that is left of Stride Rite's Massachusetts footwear manufacturing operations, though its corporate headquarters is located in the Boston suburb of Lexington. The company left Roxbury in 1981, moving its manufacturing facility to Tennessee, where cheap land and cheap labor gave it a competitive advantage in the tight children's shoe market. Stride Rite is just one example of the flight of businesses and residents from the inner city that has taken place across the country since the 1950s. The Stride Rite site is at the center of the Melnea Cass Avenue Corridor, long considered a prime location for job growth and business development. In March 1998, the Boston Water and Sewer Commission announced that it would be moving its truck depot and storage facility from South Boston to the Stride Rite site, to make way for the new convention center and related development planned for the South Boston waterfront.

Meanwhile, an innovative manufacturer of low-cost, energy-efficient housing is looking for a Boston site to build its manufacturing facility and office headquarters. The company wants to be located in the inner city, within the empowerment zone, which can provide a large part of its labor force. With over a hundred jobs to offer area residents, as well as apprenticeships, training, and plenty of other community benefits, the company presents a unique economic and environmental opportunity. However, the company is struggling to find a site and some financing. Its patience is waning.

Just eighty miles north of Boston in Manchester, New Hampshire, the Stonyfield Farm Yogurt Company is gearing up to develop the region's first ecoindustrial park, a 100-acre experiment in state-of-the-art development designed to avoid waste and pollution while maximizing economic production and performance. The brainchild of Stonyfield Farm's innovative management team and the town of Londonderry's planner, the ecoindustrial park will be among the most advanced in the world. The unemployment rate for the area is currently below 3 percent.

Community groups from Boston's inner city can often do little more than pay lip-service to these environmentally sound development opportunities. While groups like CAAP and NUSB have become experts at stopping development, they are at a loss when it comes to helping promote and implement the kind of "green," job-generating development they seek. As a result, a development vacuum is created, easily filled by undesirable projects or nothing at all, leading to the cycle of resistance and decline that characterizes too much of the land use history of America's urban communities.

What makes this challenge so significant is that environmentalists, urban policy experts, and Americans generally are coming to realize that the key to solving many of today's environmental and social problems lies in resurrecting the country's urban centers. To the extent that urban communities like Dorchester and Roxbury cannot achieve clean, job-generating economic development, not only will they continue to struggle, but the rest of the nation will suffer as population and development pressure in suburbs and rural communities continues to produce adverse environmental effects like traffic congestion, loss of habitat, and air pollution, as well as overcrowded schools and racial and economic balkanization.

This is why current environmental policy initiatives are attempting to move us closer to the ideal of democratic, sustainable communities that the landscape planner Frederick Law Olmsted described as "common place civilization" and author Jane Jacobs called an "urban village" by joining environmental and economic goals. They seek to address the widespread environmental and social degradation resulting, in large part, from shortsighted development and land use practices and urban disinvestment. For example, brownfields redevelopment (promoting the cleanup and redevelopment of contaminated land in urban centers while sparing undeveloped land in suburban and rural areas), smart growth (directing growth only to areas where the infrastructure and natural carrying capacity can accommodate it), and industrial ecology (encouraging the design of production methods and facilities that mimic the reuse, recycling, and replenishing functions of natural systems, resulting in the prevention, and not merely the control, of pollution and waste) are environmental policy initiatives, developed in the mid-1990s, aimed at protecting the environment while promoting sustainable economic development and vital communities, especially in urban areas. They seek to restore a sense of place, mixed-use development, and environmental quality that are at the core of healthy communities and look to social policy solutions that address the connections between environmental problems and economic and social issues such as urban disinvestment, racial segregation, unemployment, and civic disengagement.

These initiatives comprise an emerging model of environmental protection that moves beyond simply controlling pollution levels from smokestacks or outflow pipes and instead focuses on land use decisions and local and regional planning efforts to ensure that development occurs in accordance with environmental principles such as pollution prevention and community vision. Collectively these policies signal a revived understanding of the critical link between environmental health and economic and civic health, as well as a more comprehensive approach—a systems approach—to solving environmental problems. Each tool embraces the notion of vibrant economic activity tempered and channeled by sound environmental principles; each derives from issues of land use and community planning; each promotes a preventive, proactive approach to environmental problem solving rather than a fragmented, pollutant-by-pollutant, media-based approach.

Notwithstanding the promise of this model, existing institutions across sectors are not equipped to take advantage of these opportunities. Most community organizations and environmental groups are not used to promoting development and planning strategies, having spent much of the past thirty years advocating and litigating to stop development, as they should. The private sector, meanwhile, is still keeping its eye on the traditional bottom line, without realizing the competitive advantage afforded by environmentally responsible development and design. Lacking a strong community and regional planning tradition, government sector players are typically at a loss when it comes to implementing integrated, comprehensive planning-oriented policies.

It is for this reason that in late 1998 I founded a new kind of nonprofit environmental organization, New Ecology, Inc. (NEI), which picks up where groups like ACE leave off. It is a next-generation environmental organization aimed at helping to implement sustainable development projects in environmentally and economically hard-hit areas. NEI promotes local and regional initiatives in New England that enable a diverse set of stakeholders—from community organizations and environmental groups to municipalities and businesses—to plan, organize, and execute strategies aimed at protecting the environment while facilitating sustainable economic development and building civic capacity. With the know-how and public spirit of the third sector, and a strong connection to community-based organizations, NEI leverages its access to government and private sector resources to spearhead environmental projects that embody the core concepts of brownfields, smart growth, and industrial ecology, as well as civic environmentalism—the notion at the heart of this book that holds that social capital and environmental quality are mutually reinforcing.

NEI emphasizes environmentally sound development ecodevelopment strategies that promote the conservation of natural resources ("conservation-based development"), a sense of place ("place-based development"), and mass transit ("transit-based development"). It focuses on projects and policies that define new goals and indicators of sustainability, and link urban, suburban, and rural constituencies in the shared pursuit of livable communities. NEI takes advantage of market forces by identifying and acting on opportunities for sustainable development and planning in economically and environmentally distressed areas across New England. By serving as a catalyst for these efforts, NEI helps to establish a template of successful sustainable development models that can be replicated across the region and country, ensuring a positive feedback loop to policymakers and the private market.

Civic Environmentalism

After several years in the trenches of environmentalism, I have learned a number of lessons about the nature and extent of some of the environmental and social problems facing American society and what we can do to try to remedy them. Over time, I have come to understand that environmental action that begins at the community level (and not in corporate office towers or high-powered national public interest organizations), where ordinary people can work hand in hand with their neighbors and professionals, is, to borrow loosely from the Welsh poet Dylan Thomas, "the force that through the green fuse drives" democracy.

This book is an attempt to share my ideas and experiences in helping to build American democracy—"the land that never has been yet," in the words of Langston Hughes—from the ground up. It is an attempt to encourage action on the part of laypeople and professionals, rich and poor, whites and people of color, in communities across America to come together to reclaim democracy from the vicissitudes of history and unbridled economic growth, and, in the process, to restore environmental quality to the greatest extent possible. One of the many ironies of American life at the end of the twentieth century is that the disaffection so many Americans feel is increasingly being matched by an aroused longing for community and a strengthened value of place, for attachment to people and the environment that can restore a sense of purpose and meaning to their lives.

At the core of the book is the concept of civic environmentalism—the idea that members (stakeholders) of a particular geographic and political community—residents, businesses, government agencies, and nonprofits—should engage in planning and organizing activities to ensure a future that is environmentally healthy and economically and socially vibrant at the local and regional levels. It is based on the notion that environmental quality and economic and social health are mutually constitutive and that the protection of the environment where one lives and works is directly connected to and as important as the protection of wilderness areas or wetlands. Civic environmentalism confronts the irony that most Americans seem to care more about protecting remote natural areas than the very places they inhabit and posits the notion that we would have to spend less time worrying about protecting remote areas if we ensured that the places where people actually live are environmentally and socially healthy.

In addition, civic environmentalism denotes the admixture of a commitment to community organizing and a healthy skepticism about the promise of science and expertise to solve social ills by themselves. Civic environmentalism is about empowering a diverse set of stakeholders to work to improve and protect our natural resources (the uses and services provided by nonhuman nature) while building social capital (the capacity of communities to plan and execute socially beneficial programs) and promoting sustainable economic development (development that, to the greatest extent possible, does not occur at the expense of future generations or natural resources and promotes economic activities, at the firm and macroeconomic levels, that work as natural systems do: reusing, recycling, and replenishing and ensuring the protection of not only natural resources but the services those resources provide). Civic environmentalism is a uniquely powerful idea and practice that citizens from all walks of life can engage in their struggle to build better communities for themselves and future generations.

I did not coin the term *civic environmentalism.* Others in the environmental policy field, including DeWitt John in his 1994 book, *Civic Environmentalism: Alternatives to Regulation in States and Communities*, have used the term. John's approach focuses mainly on the role of states and municipalities in moving beyond the traditional top-down, command-and-control style of environmental regulation to a more decentralized, responsive administrative system. He argues convincingly, for instance, that states are well equipped to experiment with nonregulatory tools like education, technical assistance, and grants to address problems such as endangered ecosystems and pollution prevention.[10] He looks at the ways in which states and municipalities can assist the private sector in implementing strategies for energy efficiency and pollution prevention. He points to information generation and sharing as a key to civic environmental initiatives like designing a restoration plan for the Florida Everglades and to targeted federal regulation that complements bottom-up strategies.

John's concept of civic environmentalism is more technical than mine, stemming from his perspective as a scholar interested in public administration and models of governance beyond the traditional regulatory model. The notion of civic environmentalism embraced by *The Land That Could Be* has more to do with the civic capacity of communities to engage in effective environmental problem solving, and the relationship between the civic life of communities and environmental conditions. Although John's and my approach are similar and emphasize some common themes (though I put more stock in the overall efficacy of and need for certain traditional regulatory schemes than he does), my conception of civic environmentalism moves beyond the confines of environmental policy and administration to the arena of civic life in the broad sense.

The Progressive Policy Institute (PPI) has also espoused the idea of civic environmentalism. In a 1999 report by political scientist Marc Landy, *Civic Environmentalism in Action: A Field Guide to Regional and Local Initiatives*, the PPI has championed civic environmentalism as a "foundation for innovative, dynamic collaborations among governments, citizens, and private companies . . . better suited than command-and-control regulations to deal with certain issues, such as polluted runoff, habitat protection, and reuse of contaminated land."[11]

Through a series of case studies, the PPI demonstrates how civic environmentalism builds on the tradition of federal and state environmental regulation to create place-based solutions to environmental problems. Civic environmentalism, the report asserts, is part of the PPI's "third way philosophy," which defines a middle ground between public governance and privatization in the arena of social policy. The third way, embraced by "New Democrats" like Bill Clinton and Al Gore, seeks to blend traditional democratic concerns for the interests of lower- and middle-income Americans and public goods like environmental protection and education with market-oriented policies that stress competition and private sector leadership. The third way promotes active governmental involvement in social life, not in the form of top-down, bureaucratic authority but as an enabling mechanism, empowering citizens and local communities to take the lead in controlling their destinies.

My conception of civic environmentalism shares a lot with the PPI's but expands on and enriches this and other notions of civic environmentalism by viewing the issue primarily through the lens of civic engagement and democracy rather than environmental regulation per se. In other words, unlike prior writings on the subject, my object is to give greater weight to the civic dimension of civic environmentalism, to flesh it out and emphasize its importance, if not its priority, to the environmental dimension.

Chapter 1 describes the decline in civic participation and disaffection that are truisms of life in America at the end of the twentieth century and discusses the growing gap between the professional class and the worker and underclass, between the haves and have-nots, and between racial groups. This chapter establishes the bigger picture in terms of the existing economic and social conditions of American democracy. I discuss what democracy means (based on the ideas of people like de Tocqueville, Jefferson, John Dewey, Benjamin Barber, and Cornel West) and the disturbing picture of American democracy painted by scholars and others observers, each of whom has documented a significant amount of disaffection, loss of a sense of community, and growing economic and educational inequality among Americans.

In chapter 2, I describe how environmental or physical conditions directly reflect and are a function of social and economic conditions, borrowing from the theories of environmental historians William Cronon and Carolyn Merchant, and urban historian Dolores Hayden, among others.[12] I set forth some of the telltale features that mark the urban, suburban, and rural environment today. The chapter demonstrates how civic decline and economic and racial disparity result in pervasive negative environmental effects, such as contaminated urban land ("brownfields"), air pollution from the endless stream of cars on America's roadways, and the development of pristine rural areas. These effects corrode the fabric of American democracy across the borders of race, ethnicity, and class.

Environmental harm also endangers humans. Many environmentalists and others believe that the reckless and pervasive destruction of ecosystems will doom our own species to extinction. Whether this should come to pass, most natural scientists believe that the continued exploitation and degradation of the environment, in the United States and around the globe, will at the very least rob human societies of the many physical and biochemical functions that derive as benefits from diverse, robust ecosystems: functions such as cleaning and recirculating air and water, mitigating droughts and floods, decomposing wastes, controlling erosion, replenishing soils, pollinating crops, capturing and transporting nutrients, moderating climate fluctuations, restraining outbreaks of pestiferous species, and shielding the earth's surface from harmful ultraviolet radiation. According to noted paleontologist David Jablonski, without these beneficial natural services, "[a] lot of things are going to happen that will make this a crummier place to live—a more stressful place to live, a more difficult place to live, a less resilient place to live—before the human species is at any risk at all."[13]

My principal argument in chapter 2 is that the environment is a mirror reflecting social, political, and economic conditions. I explain that our desire for a sense of place and community and for an experiential connection to nature has been denied by these conditions, and our physical and social health undermined, and that even as we struggle to identify new symbols by which to guide our lives in the postindustrial age, we are still drawn ineluctably back to nature, however constructed or imagined. The flight to undeveloped rural areas, and the emergence of virtual communities on the Internet, are a testament to the irrepressible desire to connect ("Only connect," the author E. M. Forster enjoined), to both nature and our fellow citizens, that undergirds the practice of civic environmentalism.

In chapter 3, I describe the unique place of environmentalism and the environmental movement in American history and their relation, symbolic and literal, to American democracy. I describe the two strands of traditional environmentalism—romantic-progressive and mainstream-professional—and explain how traditional environmentalism has not been up to the task of improving American democracy because it is not itself thoroughly democratic in form or substance. Despite its noble goals and rhetoric, and its many extraordinary successes, traditional environmentalism has been narrow in its membership and agenda, comprising a homogeneous elite steeped in a romantic-progressive environmental tradition and focused on monumental natural resources to the exclusion of other issues. Traditional environmentalism has been hypocritical in its economic stance, as the wealthy elite who make up the core of traditional environmentalism have tended to disparage economic growth without proposing legitimate alternatives, thus decoupling economic challenges from environmental problem solving.

Traditional environmentalism has relied overwhelmingly on legal and policy tools to address environmental problems, dismissing the need for and rich history of grass-roots organizing and constituency building. Simply put, traditional environmentalism has lacked a meaningful and practical democratic vision, which has rendered it largely irrelevant to the day-to-day lives of most ordinary Americans.

Next, I define the fundamentals of civic environmentalism. I suggest that civic environmentalism holds great promise for restoring the health of democracy and redeeming the environment as America's great political and cultural symbol. Civic environmentalism historicizes and borrows from the best ideas in the American environmental and political tradition, beginning with Thomas Jefferson and Alexis de Tocqueville, and applies them to today's realities. It celebrates our unique environmental heritage, forged by an agrarian past, and makes that heritage relevant to an America that is now mainly urban in character.

I describe how civic environmentalism takes us from the top-down, professional problem solving of the past to the multi-stakeholder, pluralistic solution making of tomorrow. It signals a substantial shift from ex post remedies for past environmental harms to forward-looking, prevention-based strategies, and from a deficit-oriented approach—looking at what does not work—to an asset-oriented one—looking at what does. Civic environmentalism takes a systems approach to environmental problems, viewing them in the larger economic, political, and social context of which they are a part, making the connection, for instance, between civic engagement and the problem of urban brownfields. I explain the importance of local and regional governance to civic environmentalism in an age of decentralized regulatory power and diminished, though by no means extinguished, state and federal resources.

Civic environmentalism endeavors to bridge the gap between a nostalgic, idealized environmentalism grounded in a preindustrial order and a new environmentalism based in the realities of a pluralistic, multicultural, largely urban society. It is a response to the journalist Mark Dowie's assertion that, in the light of the failure of the American environmental movement to galvanize significant grass-roots support, the "real environmental movement has barely begun."[14] Civic environmentalism, this book declares, is the "fourth wave" of environmentalism Dowie envisions.

In chapters 4 through 7 I set forth a series of four case studies that illustrate how civic environmentalism is helping to improve the physical, social, and economic conditions in America's communities. Each case study is a profile in participation, planning, and action. For example, I describe the efforts of the Dudley Street Neighborhood Initiative (DSNI), one of the country's most innovative and effective community organizations, in establishing an urban agricultural vision—a modern-day, ironical, and powerful experiment in agrarianism—for the community and an action plan for achieving it. DSNI has created a model community-building process in the face of tremendous odds. I promote this and other successful efforts from Oakland, California, Routt County, Colorado, and suburban New Jersey, which, as part of a civic environmental model, establish a blueprint for other communities to adopt. These chapters explain the ways in which each of the case studies reflects civic environmental strategies and documents the challenges and failures that are part of any experience in civic environmentalism.

Each case study is a work in progress, an incomplete project. Civic environmental strategies are, in many ways, new strategies, though they borrow from a long tradition of civic action and environmental organizing. Although in some cases these projects have been underway for close to a decade, they are by nature long-term efforts, and thus their outcomes have yet to be fully achieved. As the book's title is meant to suggest, their full potential, and, in turn, the nation's, hangs in the balance. Nevertheless, they are important examples of ongoing civic environmental efforts. Accordingly, their stories deserve to be told.

By encouraging planning and organizing at the community level, civic environmentalism seeks to do nothing less than create a public discourse and dynamic social vision grounded in environmental action. It strives to inspire the kind of ethical and civic engagement that, according to political scientist Michael Sandel, self-government, civic democracy, requires. In the face of environmental harm, social disaffection, and what the writer Richard Ford calls the "strenuous [market] forces outside ourselves . . . [that] need us to want" the material goods that will never fully satisfy, civic environmentalism is attempting to restore environmental quality and a sense of place, and, in the process, rejuvenate the democratic project.[15]

1

All Things Merge into One: Making the Connection Between Environmentalism and Civic Life

Let America be America again
Let it be the dream it used to be.
Let it be the pioneer on the plain
Seeking a home where he himself is free
. . .
O, let America be America again—
The land that never has been yet—
And yet must be—the land where every man is free.
The land that's mine—the poor man's, Indian's, Negro's, ME—
Who made America,
Whose sweat and blood, whose faith and pain,
Whose hand at the foundry, whose plow in the rain,
Must bring back our might dream again.
. . .
Out of the rack and ruin of our gangster death,
The rape and rot of graft, and stealth, and lies,
We, the people, must redeem
The land, the mines, the plants, the rivers.
The mountains and the endless plain—
All, all the stretch of these great green states—
And make America again!

—Langston Hughes, "Let America Be America Again"

Democracy is not an alternative to other principles of associated life. It is the idea of community life itself. . . . It is a name for a life of free and enriching communion.[2]

—John Dewey, *The Public and Its Problems*

The Historical and Ideological Roots of Civic Environmentalism

At the heart of environmentalism is the belief in the interconnectedness of all things and life systems. At bottom, environmentalism is the expression of the inexorable faith in the wholeness of nature, the faith embodied, for instance, in Aldo Leopold's land ethic, which holds that all life is integrated in a unified, cyclical "biotic community,"[1] or in John Muir's oft-quoted observation, "When we try to pick out anything by itself, we find it hitched to everything else in the universe."[2] Perhaps the author Norman MacLean said it best: "Eventually, all things merge into one, and a river runs through it."[3] This cardinal environmental ethic underlies the basic argument of this book: that the best kind of American environmentalism fundamentally entails a holistic approach to environmental problems in that those problems and their solutions are seen as inextricably linked to social, political, and economic issues—what I collectively refer to as civic issues because each is directly associated with the quality of life of civil society, of community life in its totality.

This approach to environmentalism is akin to the systems approach to public policy developed by social scientists earlier in the twentieth century, which emphasizes an interdisciplinary focus, an appreciation of context, and an understanding of human values.[4] Systems theory originated from the work of natural scientists who studied "organized complexity"—the capability of complex systems to organize, regulate, and direct themselves. The essence of a systems approach is the understanding that the variety of components that comprise a system interact in many ways, and these components are then influenced by the new order that emerges from these interactions.

A central paradox of life in the late twentieth century is that just as modern societies have become highly specialized and compartmentalized in terms of occupations and categories of knowledge, they have increasingly come to recognize, owing to the availability of vast amounts of information and the ability to process and disseminate that information easily and rapidly, the fundamental interrelatedness of all forms of knowledge and social life. Systems thinking has become a widely embraced concept, not just in academic circles and the arena of public policy, but among community activists seeking comprehensive solutions to social problems. Civic environmentalism embodies a systems approach. It holds that in order to be effective in preventing environmental damage, environmentalists must explore issues that at first might seem unrelated or marginal but that influence environmental outcomes. All things are connected.

The connection between environmental and civic issues goes beyond theory. Our ideas about nature and the environment are themselves constructed by human values, are formed by political, social, and economic factors—by human history. As Raymond Williams has written, "the idea of nature contains, though often unnoticed, an extraordinary amount of human history."[5] The environmental historian William Cronon puts it more bluntly: "If we wish to understand the values and motivations that shape our own actions toward the natural world, if we hope for an environmentalism capable of explaining why people use and abuse the earth as they do, then the nature we study must become less natural and more cultural."[6]

Cronon, a pioneer in the field of environmental history, has documented in rich detail the ways in which human cultures shape and are shaped by ecological conditions. Cronon argues that we cannot assume that cultures, even primitive or indigenous societies, tend toward ecological stability. In explaining ecological change, Cronon urges that we see the instability of human relations with the environment as the norm. "An ecological history," he writes, "begins by assuming a dynamic and changing relationship between environment and culture, one as apt to produce contradictions as continuities. Moreover, it assumes that the interactions of the two are dialectical. The environment may initially shape the range of choices available to a people at a given moment, but the culture reshapes the environment in responding to those choices."[7] In describing the changes to New England's environment that occurred between 1650 and 1800, Cronon points to the economic forces of market capitalism, the legal regime of private property, the religious and cultural practices of puritanism, and the physical conquest and dislocation of Native Americans as the determinative factors underlying environmental change.

Another noted environmental historian, Carolyn Merchant, adopts a similar approach in explicating the causes of what she calls "ecological revolutions." Merchant argues that the course of environmental change may be understood through a description of each society's ecology, production, reproduction, and forms of consciousness.[8] By ecology, she means the relations among animals, plants, minerals, and climatic forces; by production, the extraction, processing, and exchange of resources for subsistence or profit (these are human actions with direct impact on nonhuman nature); by reproduction, the biological and social factors that influence production (for example, the size of a given human population affects the amount of resources needed to sustain that population; similarly a community's legal system or social structure influences the ways in which resources are managed and distributed); by forms of consciousness, Merchant means the modes by which societies know and explain the natural world, such as science, religion, and myth. She defines consciousness as the "totality of ones's thoughts, feelings, and impressions, [and] the awareness of one's acts and volitions."[9] According to Merchant, environmental change in American history has been dictated by the dynamic market economy developed in the eighteenth century, the European idea that civilized societies are morally superior to wild nature and indigenous cultures, and the rise of science and technology, which has enabled and justified environmental depredation.

Cronon and Merchant thus set forth the historical basis by which the environment and environmentalism should be viewed as inseparable from and embedded within a political, social, and economic matrix, a civic structure. They demonstrate that as a matter of history, environmental issues cannot be examined in isolation from the larger context from which they arise.

Beyond historical fact, however, American environmentalism is intricately tied to civic issues by the political ideology conceived by Thomas Jefferson known as agrarian republicanism. Jefferson, the American leader who, according to the journalist Brent Staples, "gave the nation its basic shape . . . and the ideas that stand as its most enduring legacy,"[10] transformed the pastoral ideal of ancient and romantic literature into a political theory based on the idea that America's landscape, its expansive environment, would itself produce a new kind of citizen who, in the words of the historian Thomas Bender, "would make the republican experiment possible."[11] Central to the concept of agrarian republicanism is the notion that good citizens live and work in the country rather than the city. Jefferson held that civic virtue, moral responsibility, and industriousness—the cornerstones of republican community and freedom—were habits born and nurtured exclusively on small farms and in rural villages, like the ones he experienced in Virginia. "Cultivators of the earth are the most valuable citizens," he declared. "They are the most vigorous, the most independent, the most virtuous, and they are tied to their country and wedded to its liberty and interests, by the most lasting bonds."[12] The very existence of fertile land, and the use of that land for agricultural purposes, Jefferson believed, enabled American republicanism to develop and thrive.

In counterpoint to Jefferson's agrarian republic were the cities of Europe. For Jefferson, as for his younger contemporary, the French aristocrat and student of politics Alexis de Tocqueville, the urban centers of the Old World were responsible for the fall of Europe from a state of preeminence and grace to one of decadence. "The mobs of great cities," he declared, "add just so much to the support of pure government, as sores to the strength of the human body."[13] Tocqueville explained America's exceptionalism in less graphic terms. The absence of great cities, he claimed, was "one of the first causes of the maintenance of republican institutions in the United States."[14] The momentous Louisiana Purchase of 1803, an effort Jefferson spearheaded, was nothing more than an attempt to ensure that, through acquisition of a vast territory stretching from the Mississippi River to the Pacific Coast, American cities would be kept invisible. Jefferson believed that the American people would "remain virtuous for many centuries, as long as they are chiefly agricultural. . . . When they get piled upon one another in large cities, as in Europe, they will become corrupt as in Europe."[15]

Having laid down the foundational political philosophy of America based on a decidedly antiurban environmental vision, and woefully limited and partial when viewed from a more modern, multicultural perspective, Jefferson forever linked Americans' political identity with their relation to the land. Of course, others after Jefferson and Tocqueville associated the vastness and beauty of the American environment, unblemished by urban sprawl, with the uniqueness and exceptionalism of the American democratic experience. In the mid-nineteenth century, the journalist Horace Greeley portrayed the great natural monuments of the West, such as the ancient redwood groves of the Pacific coast, as unique icons of democratic civilization far grander than any built monument of ancient Greece or Rome. For Greeley, these environmental monuments signified the superiority of American civilization as compared to contemporary Europe.[16]

Later, in the 1890s, the eminent American historian Frederick Jackson Turner presented his famous frontier hypothesis. This historiographical tour de force held that the frontier, which Turner defined as the meeting point of civilization and savagery, was the primary source of core democratic values such as individual freedom and responsibility. Turner argued that

"American democracy was born of no theorist's dream [but] . . . came stark and strong and full of life out of the American forest, and it gained new strength each time it touched the frontier."[17] According to Turner's view, the unsettled, undeveloped American environment made democracy possible. Never mind that Native Americans had inhabited the land later to be known as the United States for tens of thousands of years. For Turner, the historian Gordon Wood writes, "the New World that the Europeans came to in the seventeenth century was 'virgin soil,' and 'unexploited wilderness' out of which American distinctiveness was born."[18] "The existence of an area of free land, its continuous recession, and the advance of American settlement westward, explain American development," Turner argued.[19] Accordingly, he envisaged that with the closing of the frontier in the 1890s would come a crisis for American democracy.

Even the avowed city dweller Walt Whitman, writing in the mid-nineteenth century, put most of his democratic stock in the natural environment. The preeminent American poet of the city and of nature, Whitman saw the environment as a critical source of democratic freedom and a perfect antidote to Europe's feudal past. He believed America represented a brand new social order founded upon nature: "I swear there is no greatness or power that does not emulate those of the earth!/I swear there can be no theory of any account, unless it corroborates the theory of the earth."[20]

Yet with the emergence of industrialism and sprawling urban centers in America in the mid- and late-nineteenth century, Jeffersonian agrarianism and the ideology of the American environment inevitably produced a tension between inherited cultural ideals and everyday experience—the same tension Turner predicted—which led to the creation of new ideologies and a more complex, less antiurban environmental vision.[21] This vision is embodied, for example, in the work of Frederick Law Olmsted, whose urban parks, built in the last several decades of the nineteenth century, sought to integrate the freedom- and virtue-sustaining properties of rural areas with the civilizing influence of cities.

The Civics of Environmentalism

The historical and ideological connections between the environment and civic issues are rich and complex. Despite the inadequacy of America's early environmental vision, with its antiurban, nativist, and racist cast, the nation's social and environmental conditions were nevertheless joined as one. As a matter of both social change and national identity (ideology and mythology), the environment has played a central role in shaping American history, just as American society itself has dramatically altered the environment. American environmentalism is more than simply about nature; it is also about culture and the manifold civic issues culture comprises. Yet, not unlike Jefferson's own limited vision, and in part because of it, traditional environmentalism has concerned itself with a narrow set of issues, failing to embrace a larger civic or social agenda. For example, American environmentalism has typically focused exclusively on wilderness preservation and the protection of endangered species. This is not surprising, especially in the light of the ideological significance of the American environment: the American environment was conceived as the sine qua non of the American democratic experience. "To protect wilderness," Cronon writes, was for the environmental movement "in a very real sense to protect the nation's most sacred myth of origin."[22]

At the same time, American environmentalism has traditionally failed to address issues such as economic and racial equality, political participation, and economic development—civic matters that fundamentally, though sometimes indirectly, affect environmental quality. For instance, lower-income and minority communities in the United States typically have less political and economic clout than more affluent white communities. Consequently, they tend to bear the brunt of environmental harms, which often follow the path of least resistance, and they receive fewer environmental benefits, like aggressive enforcement of environmental laws or well-maintained parks or swift cleanup of contaminated land. Environmentalism, I suggest, must address the civic health of communities and not just the health of ecosystems if it is to achieve lasting results that benefit all Americans.

Indicators of civic health in a community include the strength of social networks and associations, rates of employment and poverty, the degree of participation in political and civic affairs, and the income gap between professionals and managers and wage earners. Because of the link between civic and environmental health, such civic issues also serve as potential measures of environmental quality in that they can serve as a useful proxy for more traditional environmental indicators such as the Toxic Release Inventory, the number of ozone alert days in a year, or the level of pollutants in fish tissue. They are a touchstone by which environmental progress can be measured.

Having spelled out the critical historical and ideological links between environmentalism and civic issues, I discuss the core civic matters that affect and are affected by environmental change and thus must be reckoned with by civic environmentalism. First, I set forth a working definition of democracy to ground my discussion of civic issues and environmentalism in a specific theory of democratic society. Environmentalism, like any other American social movement, must articulate a public philosophy by which its goals and strategies can be evaluated and its successes measured. A public philosophy is, as political scientist Michael Sandel explains, nothing more than the political theory implicit in our political practices and assumptions about citizenship and freedom that inform public life.[23] In the absence of an explicit and widely shared public philosophy, environmentalism may tend to drift, loosed from the principles that would otherwise direct it. Democracy, or what I call civic democracy, is that set of principles and practices that ground American environmentalism and that it must comply with in order to be true to and consistent with the aspirations of the society at large.

Next, I look at a number of issues that are indicators of civic health and explain how these indicators are related to key environmental issues.

An Environmentalist Theory of Democracy: Civic Democracy

An environmentalist theory of democracy is a civic theory of democracy. I call it "civic democracy" because it embraces the Western tradition of civic humanism or civic republicanism, which holds that the summum bonum of society is the full participation of equal citizens, including resident aliens and immigrants, in decisions that affect their lives. In its most basic form, a civic democracy provides for the regular participation of citizens in political decisions so that the making of policy and law is a shared function of the many: the ruled and the rulers are one. The political scientist Benjamin Barber calls this "strong democracy." For Barber, strong democracy, or civic democracy, is that form of government in which all of the people govern themselves in at least some public matters some of the time.[24] In a strong democracy, as in a civic democracy, individuals cannot become individuals if they do not participate in public life in that the legal rights and protections afforded individuals exist only because participatory politics have made them possible. The central virtues of Barber's strong democracy are participation, citizenship, and political activity, each based on the idea of self-governing communities. In other words, real democracy, as the social justice activist Linda Stout explains, is by definition "grass roots" in its origin and orientation.[25]

Civic democracy accepts liberalism's core concept that there exists an autonomous sphere of public life where people govern themselves and treat each other fairly without promoting any particular person's or association's good; then it ratchets it up a few notches. In a civic democracy, equals come together voluntarily to promote the diverse interests of the group as a whole. As the historian Christopher Lasch describes, "self-governing communities, not individuals, are the basic units of democratic society."[26] For the pragmatist John Dewey, "democracy is not an alternative to other principles of associated life. It is the idea of community life itself. . . . It is a name for a life of free and enriching communion."[27] Democracy is what results when all people have the opportunity to develop and use their capacities to the fullest extent possible; it fails when the privileged "shut out some from the conditions which direct and evoke their capacities."[28] For Dewey, democracy empowers people to work together, initiate action, experiment, learn facts, and solve problems.

In the spirit of Jefferson and Dewey, the philosopher Richard Rorty explains, institutions in a civic democracy must be viewed as experiments in "cooperation rather than as attempts to embody a universal and ahistorical order."[29] As Rorty describes, key measures of success in a civic democracy include the openness and responsiveness of its decision-making procedures and its sensitivity to outsiders and those who are suffering.

Michael Sandel expands on the idea of civic democracy by explaining that civic republicanism promotes liberty not simply through the fair procedures that liberalism exclusively supports, but through the act of sharing in self-government and deliberating with fellow citizens about the common good. Democratic freedom, Sandel argues, requires more than just due process; it demands "multiple sites" of civic formation where a sense of community and common purpose can be inspired.[30] These sites are the essence of a civic democracy because they are where citizens convene, listen and talk to each other, plan, and make decisions; they are the bedrock of civil society and civic engagement.

Civil society is the musculature of democracy. It is the schools, workplaces, trade unions, churches and synagogues, and associations that, in addition to local government and municipal institutions, are the public spaces and activities that "gather citizens together, enable them to interpret their conditions, and cultivate solidarity and civic engagement."[31] Frederick Law Olmsted called civil society "common place civilization."[32] Civil society also includes the neighborhood pubs, bookstores, and coffeehouses that Ray Oldenburg describes as the "third places" of society—the places where people can meet as equals, without regard to race, ethnicity, class, sexual orientation, or national origin. They are the places that encourage, in the words of Jane Jacobs, "casual public trust" and link neighborhoods to the larger, more impersonal world. In turn, civil society consolidates democratic life by ensuring successful outcomes in education, poverty, employment, crime and drug abuse, and health.

The strength of civil society is itself a function of yet another element of civic democracy: social capital. Social capital, a term first employed by Jacobs in her landmark book, *The Death and Life of Great American Cities*, denotes the networks, norms, and social trust that facilitate coordination and cooperation among people for mutual benefit. These rudimentary features of social organization broaden one's sense of self, developing, as the political scientist Robert Putnam explains, the "I" into "We."[33] Interpersonal and interorganizational networks foster "sturdy" norms of reciprocity and promote social trust; they facilitate cooperation and reduce incentives for opportunism. "Life is easier," Putnam declares, "in a community blessed with a substantial stock of social capital."[34]

Civic democracy thus rests on what the philosopher Cornel West describes as the ability of ordinary people, and especially the dispossessed, to participate meaningfully in the decision-making procedures of institutions that fundamentally regulate their lives, and to engage daily in discourse, planning, and coordinated action with their fellow citizens to promote the pubic interest.[35] It is the ongoing process of democratic discussion in which citizens engage in public thinking and political judgment aimed at building a community of common purpose and vision.

Civic democracy is more than just community participation and conversation; it is rooted in a place, a physical environment conducive to collective action and community building. "Communities . . . are as much results as they are causes of their own environments," cultural theorists Laurie Anne Whitt and Jennifer Daryl Slack explain, and the relation between a particular community and its environment 'is not simply one of interaction of internal and external factors, but of a dialectical development' . . . of community and environment in response to one another."[36] In a civic democracy, place and community are mutually constitutive and reinforcing.

For Aristotle, the polis—the small, geographic unit of the city-state whose boundaries were never more than a day's walk away—was the physical setting most beneficial to democracy. For Jefferson and Tocqueville, it was the wards and townships of early America, the small-scale communities that permitted direct political participation, albeit by landed white males only. For Benjamin Barber and Jane Jacobs, real democracy thrives only in a well-designed, well-planned community that ensures access for people to local institutions, whether a park, community center, or city hall, so they can be involved in at least some of the decisions that affect the community as a whole. Thus, Barber exhorts, "Strong democratic community will have to find new forms of physical dwelling if it is to thrive in large cities or suburban landscapes."[37]

One of the great travesties of the urban renewal programs of the 1950s and 1960s, Robert Putnam explains, was that the slum clearance policy renovated physical infrastructure "at a very high cost to existing social capital."[38] From the standpoint of civic democracy, the public housing developments and other major building projects of the urban renewal era were a dismal failure; they shunted groups of people off from the larger community through imposing an impersonal architecture and crude planning. Place plays an essential role in a civic democracy because it is the physical infrastructure out of which neighborhoods, associations, and local politics develop. Perhaps this is one reason that Tip O'Neil, the avuncular career Massachusetts congressman, always insisted that "all politics is local."

Civic democracy is thus the amalgam of civil society, social capital, and the local environment. It is the framework within which all aspects of society must be evaluated and the yardstick by which they must be measured.

Core Indicators of Civic Health

With a public philosophy of civic environmentalism at hand, we can now examine some of the social, political, and economic factors that inhibit or enable the achievement of a civic democracy and, in turn, a healthy, sustainable environment. Each factor has a corresponding set of environmental effects, all of which must be accounted for in a civic environmental program, within the framework of civic democratic principles. Such factors are indicators of civic health that serve as the main feedback loop to local planning and policy decisions and help define the problems or opportunities that community members, professionals, policymakers, and politicians seek to address.

Environmental success can be measured, at least in part, by viewing the overall state of civic democracy through a lens of a series of indicators of civic health. As in a system, civic and environmental conditions interact in often complex and subtle ways. Therefore, environmentalists must engage civic issues when it comes to devising solutions to environmental problems. This is the basic premise of civic environmentalism. In chapter 2, I spell out the environmental issues that correspond to the civic indicators described below.

Social Capital A recent study on social health conducted by the Fordham University Institute for Innovation in Social Policy found that "of the eight worst years since 1970, six have been in this decade [the 1990s]. The social health of the nation has not kept up with the recovery of the economy."[39] The study looked at sixteen indicators of social health, including the percentage of children and elderly in poverty, unemployment rates, homicides, health insurance coverage, access to affordable housing, and infant mortality. The report noted that the United States is the only industrialized nation that does not officially monitor overall social progress.

Robert Putnam's research corroborates the Fordham study. He claims "there is striking evidence . . . that the vibrancy of American civil society has notably declined over the past several decades." The number of Americans who reported that in the past year they attended a public meeting on municipal or school affairs dropped from 22 percent in 1973 to 13 percent in 1993. "By almost every measure," Putnam declares, "Americans' direct engagement in politics and government has fallen steadily and sharply over the last generation."[40]

Documenting membership in civic organizations like the Red Cross, Lions, Elks, League of Women Voters, Boy Scouts, and bowling leagues, Putnam finds that despite steady increases throughout most of the twentieth century, many major civic organizations—what Putnam calls "secondary associations"—"have experienced a sudden, substantial, and nearly simultaneous decline in membership over the last decade or two."[41] While mass membership organizations such as the Sierra Club, the National Organization for Women, and the American Association of Retired Persons remain strong, Putnam points out that these are very different from the civic associations on which the development of social capital depends. These organizations—what he calls "tertiary organizations"—do not foster interpersonal ties or social cohesion. Rather, membership is tied to a cause or issue. Even the growing prominence of nonprofit organizations that are not secondary associations, so-called third-sector organizations, does not affect the amount of social capital in a community. They are just not the same as bowling leagues, Putnam notes. Informal kinds of social capital, such as family ties, neighborliness, and trust, are also lacking.

Putnam points to four principal factors responsible for the decline in social capital, the feedstock of civic democracy: the movement of women from civic associations into the workforce; the mobility of workers in modern society, which means less residential stability and rootedness in a place; the recent changes in economic scale, from the mom-and-pop grocery to the superstore, from the local to global market, which have undermined the material and physical basis for civic engagement; and the technological transformation of leisure from group activities, such as bowling leagues and community theater, to private, individualized pursuits like television and VCRs. Each of these factors, Putnam argues, has contributed to the decline in social capital that marks the last two decades of the twentieth century. Nevertheless, he cautions, we must not view this decline as purely negative or indiscriminately romanticize small town, middle-class life of the past. For example, likely salutary effects of the decline in social capital are less intolerance and overt discrimination and the advancement of women in the workplace.[42]

Nonetheless, despite significant improvement in the quality of life for many Americans over recent decades, something fundamental has been lost. As the sociologist Robert Bellah suggests, American democracy in the postindustrial age is characterized largely by selfishness and apathy, behaviors that are the antithesis of what Tocqueville called the "habits of the heart" essential to maintaining the American democratic system.[43] Even the salutary tolerance and live-and-let-live attitude embraced by much of the middle class, warns the sociologist Alan Wolfe in his book *One Nation, After All*, can be taken as evidence that most Americans do not care what their neighbors do or what happens in their communities.[44] The diminution of social capital has resulted in a social vacuum, seemingly to be filled not by innovative surrogates but by the vices born of excessive individualism and privatization.

Political Participation Closely linked to the decline of social capital are the alarmingly low rates of political participation in American society. Political participation is a strong indicator of civic health because it demonstrates unambiguously the willingness of citizens to engage in the political process and thereby to affect public policy outcomes. In a civic democracy, the quality of life of individuals is de facto a function of the quality and quantity of public participation in every facet of social life, including politics. As Benjamin Barber explains, "Without participating in the common life that defines them and the decisionmaking that shapes their social habitat, people can't become individuals. . . . Our most deeply cherished values are all gifts of law and of the politics that make law possible."[45]

That American society has become more democratic over time cannot be disputed. Since the end of the Civil War and through much struggle and hardship, new groups such as blacks and women have won the franchise; schemes to deny voting rights like poll taxes, literacy tests, and malapportioned voting districts have been abolished; and greater numbers of minorities have gained political office.

Yet as many observers of today's political scene have noted, something is terribly wrong with American politics. In the age of global democratization, it is fair to say that the world's leading democracy suffers from too little democracy. Since the end of World War II, the mean voter turnout rate has hovered around 50 percent, lower than every other noncompulsory democracy in the West.[46] Less than one-third of voters who are not upper-income professionals participate in American elections.[47] With each election year, the American electorate seems to fall to new lows in voter turnout.

In a recent Boston City Council election, only 16 percent of the residents of the populous Roxbury neighborhood bothered to vote. Such statistics are not unique.

As recent debates about campaign finance reform reveal, money is a major factor in American elections. The financing of political campaigns by wealthy individual donors or candidates themselves has exerted an undue influence on the political process and inhibited participation by all but the wealthy and powerful. As cultural critic Michael Lind explains, "Campaign financing is by far the most important mechanism for overclass influence in government."[48]

Voting rights attorney John Bonifaz, director of the National Voting Rights Institute, calls this mechanism the "wealth primary." Because the wealth primary system is controlled by "large blocks of private wealth," voters from lower-income or working-class backgrounds are effectively denied "the opportunity to affect the political programs and positions of candidates in the race. Citizens without money to give are totally excluded from participation in the wealth primary."[49] The bottom line of the wealth primary, and thus of American politics, is that the candidate with the most money wins.

Money in politics has thus had a profoundly corrosive effect on Americans' willingness to vote and, consequently, to influence the decision making that fundamentally affects their lives. That social capital should be declining at the same time that political participation wanes is to be expected. Social capital and political participation are mutually reinforcing elements of civic democracy. Without them, civic democracy cannot be sustained.

Racial Equality Despite the heroic efforts since midcentury, full racial integration and equality still elude American society. Most white Americans, and especially the elite, inhabit communities composed almost exclusively of people who look and act as they do. Since the racially turbulent decade of the 1960s, when whites began fleeing mixed-income urban neighborhoods for the security of the suburbs, America has remained racially polarized. In most major American cities, the central neighborhoods and public schools are still populated largely by people of color; the outlying suburban communities are overwhelmingly white. Nearly 60 percent of all African

Americans and close to half of the Hispanic and Asian populations live in America's central cities; only a quarter of the white population lives in urban centers.[50] Despite real gains in the diversification of the workplace and higher education over the past three decades, residential segregation, brought about by both discrimination (outright and aversive) and lack of economic power, persists.

The economic gap between whites and nonwhites is startling. In the nation's one hundred largest cities, where most black Americans and other minorities live, one in seven census tracts is at least 40 percent poor.[51] Twenty-six percent of all black families and nearly 42 percent of black children are trapped in poverty.[52] Rates of poverty among Latinos and Native Americans are equally disturbing. For example, the Pine Ridge Reservation in South Dakota, home to the Oglala Lakota tribe, is consistently among the poorest communities in the nation. "The poorest of the poor—by far—," the author Peter Mathiessen laments, "are the Indian people."[53] Notwithstanding the recent call for an end to affirmative action, based on the belief that blacks, Latinos, and other minorities have achieved a level playing field on which to compete with whites, most social scientists agree that racial inequality is as problematic today as it was fifty years ago. The sociologist Orlando Patterson, who is quick to point out that blacks and other minorities have made extraordinary progress in overcoming racism and integrating society, blames pervasive income inequality among racial groups as the real threat to American democracy while acknowledging that 20 percent of whites are still at least mildly racist.[54]

In their pathbreaking study, *Black Wealth/White Wealth: A New Perspective on Racial Inequality*, sociologists Melvin Oliver and Thomas Shapiro report that African Americans have not shared equally in the nation's overall prosperity. White Americans, they document, possess nearly twelve times as much median net worth as blacks while black households retain no net financial assets as compared to approximately $7,000 per white household.[55]

Even the Internet, the much-touted tool of digital democracy, is afflicted by racial inequality. A recent study by Vanderbilt University professor Donna Hoffman showed that black Americans are far less likely to use the information highway than whites. In households with annual incomes below $40,000, whites were six times as likely as blacks to have used the

World Wide Web. Lower-income white households were also twice as likely to own a computer as black households. Among households of all income levels, 44.3 percent of whites own a home computer compared with only 29 percent of blacks. While 73 percent of white high school and college students had access to a computer at home, only 32 percent of black students had access to one.[56] In the light of the growing importance of the Internet for communication, business, and education, the study found, the racial divide on the Internet might well exacerbate existing social and economic disparities among racial groups

The eminent sociologist William Julius Wilson, in his 1996 book, *When Work Disappears: The World of the New Urban Poor*, declares that "for the first time in the twentieth century most adults in many inner-city ghetto neighborhoods are not working in a typical week."[57] Pointing to inequities in the society at large to explain the economic and social marginality of inner-city residents, Wilson sees the decline in formal and informal community institutions (churches, political parties, block associations, parent-teacher organizations) and social networks (workers, family, friends) as largely responsible for the economic crisis in inner-city neighborhoods. Any explanation of the plight of the urban poor, Wilson argues, must account not simply for race but culture and social psychology as well. Nonetheless, he suggests, urban poverty is still mainly associated with racial minorities and must be dealt with accordingly.

Thus, in spite of significant progress toward a racially integrated society, pernicious barriers remain. Physical segregation between whites and people of color and substantial economic disparities reveal that we have yet to achieve the kind of democracy that most of us claim to want.

Socioeconomic Equality Linked to the problem of persistent racial inequality in America is the ever-growing gap between wealthy and less affluent Americans. To the extent that political power derives in large part from wealth and economic status, as the wealth primary example shows, the size of the disparity between income groups is a good measure of the relative success or failure of civic democracy in providing a level playing field on which citizens can participate as equals in decision making. Notwithstanding the unparalleled economic prosperity of the past several years, with historically low unemployment rates and a skyrocketing stock market, there remains a nagging sense among many Americans that the only real beneficiaries of the strong economy are those who are already financially secure.

The data bear this out. In 1997 the income of the top 1 percent of Americans, or about 2.6 million people, equaled that of the bottom one-third of the population, or 88 million people.[58] Income inequality was greater in the 1990s than it has been since the era of the Great Depression in the 1930s. Today the wealthiest 20 percent of Americans control over 50 percent of wealth in the United States.[59] Moreover, the average American chief executive officer makes roughly 109 times as much as the average American blue-collar worker. In the 1980s, worker pay increased 53 percent; CEO compensation soared a whopping 212 percent.[60]

The rift between the haves and have-nots is driven by two main forces: professionalism and global capitalism. Professionalism is a modern phenomenon that arose out of the nineteenth century, when industrialism and science created new economic conditions. Whereas in preindustrialized economies, producers and consumers were essentially one and the same, as in colonial America, increasingly complex industrialized economies, made possible in the 1800s by rapid advances in science and technology, eventually required a cadre of educated individuals whose job it was to protect the autonomy of skilled producers while ensuring the quality of services and public safety.[61] By definition, professionals are credentialed or licensed experts whose power and legitimacy derive from their ability to render objective opinions based on rationality and scientific knowledge. Unlike purely economic actors motivated by profit and self-interest, professionals are supposed to be disinterested and quality driven. Their power is based not on labor or capital, the traditional bases of capitalist wealth, but on knowledge. The nontransferability of that knowledge and expertise is meant to ensure that the well educated do not make exaggerated claims to power.

With the rise of universities and professional journals early in the 1900s, and the parallel ascendancy of progressive ideology that stressed the importance of expertise and rational decision making, the number of professionals also increased. In 1890, there were fewer than 1 million professionals in the United States. By 1986, professionals accounted for over 13 percent of the labor force.[62] As the cultural critic Louis Menand states, the rise of professionalism has engendered "resentment on the part of people whose employment is relatively insecure and whose incomes are stagnant or in decline. Professionalism has started to look like a racket."[63]

The success of professionalism has come at a high cost to democracy. "Professionalism erodes the right of those not certified as experts, bringing its own threat to democracy and equality," sociologists Charles Derber, William Schwartz, and Yale Magrass claim. "The shadow side of professionalism is the creation of a new dispossessed majority: the uncredentialed."[64] Professionals are in fact a kind of "overclass," in the words of Michael Lind, who spend most of their waking hours as lawyers, real estate developers, investment bankers, software designers, and other credentialed players as intent on preserving their professional fiefs as they are on "perpetuating the conditions that make it possible for them to ignore the views of ordinary people."[65] Professionals live out their days largely isolated from ordinary Americans, in office towers or exclusive suburbs, preoccupied with the abstract concepts and symbols that are their stock and trade: stock market quotations, the visual images of Hollywood and Madison Avenue, computer software, and the like. Stronger than ever, today's professional class, fortified by the information economy, is "switched-on [and] plugged-in . . . with a seemingly insatiable appetite for talking to itself."[66]

Professionals tend to be migratory and cosmopolitan in their disposition, hopping from metropolis to metropolis with barely an opportunity along the way to become attached to a particular place. According to Christopher Lasch, professionals possess a "tourist's view of the world—not a perspective likely to encourage a passionate devotion to democracy."[67] Professionals are thus detached from the destiny of wage-earning Americans, their fate hitched to the forces of modern capitalism and their lives ensconced within exclusive work, residential, and leisure environments.

Another leading contributor to the growing gap between haves and have-nots is global capitalism. Linked to the rise of the professional-managerial class in the last several decades and the recent spread of market-based economies to formerly communist countries throughout the world, global capitalism is the economic juggernaut of the end of the millennium. Aided by major international trade agreements like the North American Free Trade Agreement and the General Agreement on Tariffs and Trade, multinational firms from the United States and other industrialized economies have seen their profits soar as they have tapped into massive new markets and cheaper sources of labor. From high tech to financial and legal services to manufacturing and retail, the manifold sectors of the American economy, and especially upper-level management and professionals, have benefited significantly from the globalization of commerce and capitalism.

Bolstered by their international reach, the interests and aims of American multinationals have, according to Cornel West, "disproportionately" shaped American society, resulting in "vast disparities in resources, wealth and income."[68] Global capitalism, like capitalism generally, has very different core values from democracy. As economist Lester Thurow writes, "Democracy believes in radical equality. . . . Capitalism believes in radical inequality."[69] These words echo those of political scientist Charles Lindblom, who declared in his 1977 book, *Politics and Markets*, "The large private corporation fits oddly into democratic theory and vision. Indeed, it does not fit."[70] Big, global businesses, it appears, increasingly regulate the lives and livelihoods of most American workers, whose fate is now subject to control by competition from foreign workers whose wages and work environment give them a competitive advantage in the race-to-the-bottom world of global commerce. Many American workers, their wages having stagnated over two decades, are at constant risk of losing their jobs as corporate executives continue to gravitate to the cheaper labor markets of undeveloped foreign countries.

In sum, the growing gap between rich and poor is a function of the significant physical and social separation of professionals and corporate executives from ordinary Americans. "The elite," law professor Charles Reich observes, "live in a different country than the rest of America."[71] With a direct pipeline to the vast profits derived from a strong stock market and ever expanding global markets, the wealthiest Americans enjoy a social reality that is increasingly independent of the goings on on Main Street, the inner city, or impoverished rural communities.

In addition, more often than not, professionals and corporate managers tend to act as if only one thing matters: greater accumulation of wealth. Most do not even purport to serve the public interest or take responsibility for the larger social problems that surround them and to which they contribute, whether it is destruction of natural resources in the name of economic growth, or laying off hundreds of workers because of global competition. The yawning gulf separating the haves and have-nots is thus like a great vacuum, sucking up Americans' public spirit and willingness to solve social problems.

Public Investment and Privatization Civic democracy, by virtue of the priority it gives to participatory politics and civic engagement, rests on a social and physical infrastructure capitalized largely by public investment. In the light of the free market's tendency to pursue short-term, individual gains over longer-term, public benefits, social investment in public goods like law enforcement, education, research and training, transportation, housing, and the environment are essential to ensuring civic health. Social investment in education and other public goods historically has served to reconcile capitalism's values with democracy's, as exemplified by the reforms of the New Deal era. Without such investment, civic democracy inevitably languishes, eventually succumbing to a minimalist liberal state where individuals pursue their own ends without regard for social consequences and where the economically fit dominate the economically disadvantaged, as happened in the nineteenth-century Gilded Age, when robber barons made their fortunes literally on the backs of impoverished immigrants and migrant workers.

At the heart of privatization and social disinvestment is liberal political theory and the private property system that is its most powerful instrument. Liberalism lends itself to privatization; its core precept is the priority of the individual to the state and the notion that the state must remain neutral among competing conceptions of the good held by its citizens. Liberalism sanctifies individual choice and value, allowing each person to pursue her particular ends relatively unfettered and oblivious to the social consequences of her actions. The private is thus superior to the public; the individual, to amend Protagoras's maxim, is the measure of all things.

Not that such a political philosophy has no benefits for democratic life and civil liberties. Minimal state encroachment preserves an exclusive realm of private affairs that can enhance individual freedom by providing opportunities for individuals to lead their lives according to their own preferences and principles. Nevertheless, liberalism's thrust is inevitably toward the private, typically at the expense of public life.

Liberal political theory embraces private property as the key to individual liberty. The link between liberalism and property was forged by the father of modern liberalism, the English philosopher John Locke, who held that the appropriation and use of land defines property and gives rise to rights in it.[72] In other words, property and property rights do not exist unless and until property is appropriated by individuals. The German philosopher G. W. F. Hegel went a step further. He maintained that all people have a basic right to possess property; without it, they are not capable of full ethical development.[73]

Liberalism rests on the ability of individuals to own and use property, with minimal regulation by the state. In a liberal society, property ownership enables individuals to achieve their full potential. On a more practical level, property rights provide owners and potential owners with the security they need to invest in their property and to engage in exchange. These are the foundational elements of a capitalist system that create the economic framework within which a liberal society can operate.

Liberalism and private property are the necessary precursors to capitalism and the free market. The pursuit of profit and economic growth are made possible by liberalism's minimalist state and the ability of individuals to amass great quantities of property or capital. The American public, through legislation and the common law, has always played a part in controlling the market for the sake of the public interest. From constitutional measures like the interstate commerce clause and the civil rights amendments, to the imposition of income taxes, to the regulation of securities and banks, Americans have wielded considerable authority over the market. Even during the heyday of the nineteenth-century American economy—the so-called laissez-faire era—market forces were subject to legislative control. The eminent domain power that railroad companies and utilities exercised in the mid-1800s, for example, was the result of legislative action devolving this public right to private actors. Still, there was, and is, no other country in the world where capitalism has enjoyed so much free rein as in the United States. Is it any wonder that among our most celebrated examples of public architecture are our banks?

In the last twenty-five years, social investments have been cut in half.[74] In the 1980s, when the Republican mantra of less government and more free market first took hold of Americans' political imagination, federal and state spending for law enforcement declined by 42 percent, for education and training by 40 percent, for transportation infrastructure by 32 percent.[75] Since the mid-1970s, many of the social programs and worker protections afforded by the New Deal and the American labor movement have been dissolved. At the same time, more and more public matters are being relegated to the private sector, among them education, welfare, and transportation.

According to Alan Wolfe, most middle-class Americans acknowledge that many social problems exist yet are disinclined to address them. Most Americans, Wolfe bemoans, "refuse to accept the responsibilities of national citizenship. They seem to want the benefits of being American without the obligations of paying taxes or paying attention." As in the halcyon era of the 1950s, Americans today appear to prefer the "pleasures of private success to the rewards and frustrations of public involvement."[76]

Meanwhile, affluent Americans have effectively seceded altogether from the civic life of their communities. Through private schools and country clubs, private security services, gated communities, and the like, the wealthy have "[bought] their way out of reliance on public services," eroding the "formative, civic resources of American life."[77] Moreover, as the wealthy and poor have grown further apart economically and physically, affluent Americans' sense of shared fate has diminished along with their willingness to invest, through taxes and charitable contributions, in the education and welfare of their less fortunate fellow citizens. The privatization of American culture has resulted in a politics that reformulates public goods in terms of private advantage and fails to nurture a sense of genuine public interest or inspire affirmative community action in the pursuit of common goals.

With the advent of fax machines, the Internet, modems, and cellular phones, privatization has been buttressed and accelerated by a regime of products that allows individuals to work full time in the privacy of their home or automobile. Although some, like the writer Michael Pollan, try to explain that these digital devices enable workers to engage in more civic activities than they normally would because, having been cooped up at home all day, they are more inclined to attend community meetings at night, there is little doubt that the digital age increasingly has isolated individuals, who seek community not with their neighbors but with the virtual communities of the Internet.[78] Just as new technologies have empowered and enriched countless professionals and businesspeople, they have accelerated the disabling of public life by consuming the time once available for face-to-face, communal interaction and replacing it with on-demand, on-line correspondence, as placeless as it often is anonymous and transient.

Increased privatization has come at the expense of public institutions, the places where citizens can meet as equals, and has insulated the rich from the lives of ordinary Americans. The noted economist John Kenneth Galbraith predicted as much over forty years ago in his book, *The Affluent Society.* Galbraith warned that private sector wealth would inevitably undermine the viability of the public sector, resulting in "public poverty" reflected in inadequate public services, environmental degradation, and failing public schools.[79] Today, the power of the market has penetrated almost every sector of social life, enshrining money, private property, and technology, while devaluing basic public goods like education, health care, and physical infrastructure such as parks and mass transit. "Civic life," Christopher Lasch explains, "requires settings in which people meet as equals, without regard to race, class, or national origin."[80] Such settings, it appears, are no longer open to the public.

2

The Environmental Consequences of Civic Decline

To build a better motor we tap the uttermost power of the human brain; to build a better countryside, we throw dice.
—Aldo Leopold, *A Sand County Almanac*

In nature, nothing takes place in isolation. . . . Each conquest takes its revenge on us.
—Frederich Engels, *Dialectics of Nature*

The Relationship Between Civic Health and Environmental Quality

Judging by the sample of indicators discussed in chapter 1, the civic health of our democracy is in need of considerable repair. Across the spectrum of issues that reflect the core of our democratic values and aspirations, Americans have in many ways failed to live up to our ambitious commitments. Yet as public intellectuals like Orlando Patterson are quick to remind us, we have achieved a lot in the way of social progress in the last half-century: universal suffrage, legal protection of civil rights and liberties, and a strong measure of tolerance. Though we have abandoned a certain amount of innocence and idealism, we have adopted many pragmatic, responsible solutions to a variety of social problems; and although the traditional family and community structure have withered, institutions such as universities, the professions, and corporations have open their doors to women and minorities. To borrow a line from the poet Lord Tennyson, "Though much is taken, much abides." Our development as a society has not been marked exclusively by progress or decline; it has not been linear but more cyclical.

Nonetheless, the current social condition of civic democracy leaves a lot to be desired. The decline of social capital, lack of significant political participation, persistent racial and economic inequalities, and increased privatization and commercialization are facts of American life at the end of the twentieth century. They have become annoying truisms that seem to reinforce each other in a degenerative cycle of social malaise.

This is not all. Our atrophied civic life is also reflected in and affected by the physical condition of America's communities and landscape. That civics expresses itself in physical form is nothing new. Civic value has long been a concern of great architects, whose buildings are intended to enhance civic life by virtue of their beauty and scale. As the nineteenth-century British art historian John Ruskin explained, "Architecture is the art which so disposes and adorns the edifices raised by man for whatsoever uses, that the sight of them contributes to his mental health, power, and pleasure."[1] Whether exemplified in the arches and basilicas of ancient Rome or, later, in the "white city" buildings of the famous Columbian Exposition in Chicago in 1893, civic architecture has historically been a tool by which communities, from villages to nation-states, express their political and cultural ideals and appeal to their citizens' noblest sensibilities.

Civic expression goes beyond architecture to land use and the environment itself. The material and physical ways in which our communities are developed and the social, cultural, and economic activities—the civic activities—that underlie that development result in certain discrete, though often intertwined, environmental effects. Contaminated urban land, suburban sprawl, polluted rivers, drained wetlands, regional smog, acid rain, clear-cutting, and endangered species: these are some of the adverse and interrelated physical effects of development that ultimately are a reflection of the civic health and consciousness of communities.

The social capital in a community acts on and is itself acted on by the natural resources of that community—what businessman Paul Hawken calls "natural capital." Natural capital comprises the resources we use, both nonrenewable (oil, coal, metal ore) and renewable (forests, fisheries, grasslands), and the services those resources provide (forest cover, parks, topsoil, clean air, rainfall, waste processing, medicines, erosion control).[2] In communities endowed with a substantial amount of social capital, we tend to see an equivalent amount of natural capital; in those without a good supply of social capital, we tend to find depleted and degraded reserves of natural capital. Social capital and natural capital should not be seen as a function of the relative material wealth of communities, though economic power can increase the capacity of communities to create and manage these forms of capital or, alternatively, to harm and deplete them. Rather, they denote the civic and environmental resources that people cultivate, maintain, and protect, regardless of their socioeconomic status.

The civic elements in a community—the education system, the rates of poverty and unemployment, the level of political participation—and the physical environment are thus reciprocating conditions. As environmental historians like Carolyn Merchant and Dolores Hayden explain, the human forces of production, reproduction, and consciousness within the American liberal-capitalist system shape and reshape the physical environment, while the environment itself influences and constrains those same forces.

At its most basic level, environmental health contributes to the physical well-being of human communities. The physical and biochemical functions performed by healthy ecosystems enhance the quality of human life by providing essential services such as cleaning and recirculating air and water, mitigating droughts and floods, controlling erosion, moderating temperatures, and protecting the earth's surface from harmful radiation. Although without these functions, humans might well survive in what author David Quammen calls a "Planet of Weeds," comprising only the hardiest, most opportunistic species, such as crows, kudzu, and cockroaches, diverse, robust ecosystems ensure a relatively safe, peaceful habitat for human communities.[3]

In his seminal work, *The Unsettling of America*, Wendell Berry describes the relation between the environment and community in terms that go beyond the merely physical:

> We and our country create one another, depend on one another, are literally part of one another; that our land passes in and out of our bodies just as our bodies pass in and out of our land; that as we and our land are part of one another, so all who are living as neighbors here, human and plant and animal, are part of one another, and so cannot possibly flourish alone; that, therefore, our culture must be our response to our place, our culture and our place are images of each other and inseparable from each other, and so neither can be better than the other.[4]

As Berry suggests, our sense of place and experience of nature are bound up in the relation between civic and environmental health. Much has been written about the importance of place and nature to the ability of citizens to function effectively in a community. The eminent urban historian Lewis

Mumford was among the first to equate place and citizenship. Mumford claimed that the newly emergent city-states of ancient Greece produced a new kind of person—the citizen. The geographic and political place, embodied in the city-state, he pronounced, made the citizen possible.[5]

In a world of diminishing spatial barriers to movement and communication, historian Dolores Hayden states, "place-bound identities," such as those cultivated in Socrates' Athens, "become more, not less important" to ensuring the physical and social health of local communities.[6] Ecologist and philosopher David Abram argues that in the seemingly unlimited and global, technologically mediated world of today, we must "reinhabit" the local places and regions where we can still have direct sensory interactions with the environment. By "apprenticing oneself to one's place and region," Abram claims, we can begin to restore damaged habitats and the balance between civilization and nature. "Lacking all sacredness, stripped of all spiritual significance," Abram laments, "the air is today little more than a conveniently forgotten dump site for a host of gaseous effluents and industrial pollutants."[7] For Abram, the sensuous experience of place is the key to environmental problem solving.

Industrialization and technology have disturbed the fragile balance between people and places by generating a "structure of place." Good and bad experiences are now segregated into different places, such as slums and suburbs. Modern cities and towns, claims the sociologist E. V. Walter, are filled with barriers, both material and intangible, "which conceal or deny that people with different social identities defined by class or by ethnic characteristics dwell in the same town. This segregation saps the vitality of places."[8]

Moreover, without a sense of place, people are rootless and want to drift. Notwithstanding the benefits that the information age has brought in the form of greater social awareness and more tolerance, it has also diminished our sense of place and time and, as a consequence, has caused us to lose a sense of history and vision of a collective future.

Place can be an enabling factor in the lives of communities. As the writer Tony Hiss explains, "We all react, consciously and unconsciously, to the places where we live and work. . . . These places have an impact on our sense of self, our sense of safety, the kind of work we get done, the ways we interact with other people, even our ability to function as citizens in a democracy. In short, the places where we spend our time affect the people we are and can become."[9] Similarly, place can nurture public memory, the sense of civic identity, which empowers citizens and inspires them to contribute to civic life. The "power of place," Dolores Hayden describes, "is the power of ordinary . . . landscapes to nurture citizens' public memory, to encompass shared time in the form of shared territory."[10]

Part and parcel of a sense of place, the experience of nature contributes to the well-being of communities by providing restorative opportunities amid the discontinuities and stress of everyday modern life. The psychologists Rachel and Stephen Kaplan have found that diverse and healthy natural settings in urban and rural contexts enhance human cognitive and psychological functions, across demographic and socioeconomic categories.[11] Further, they discovered that wilderness areas are not the only settings that offer restorative experiences. What is required for such experiences, the Kaplans conclude, is only that the setting provide a sense of being away, scale, fascination, and compatibility with the surrounding physical and social environment.

Many American cities and corporations, especially in the West, have incorporated such findings in their planning. The city of Portland, Oregon, and corporations such as Hewlett-Packard, Intel, and Hyundai have approached their physical infrastructure with an eye toward integrating it with undeveloped natural settings. "If anything has defined the quest to create something different in the emerging mega-cities of the West," the journalist Timothy Egan writes, "it is the idea of blending urban life with the outdoors—be it the desert, the Rocky Mountains, or the damp woods of the Pacific Northwest."[12]

The relation between the environment and civic life is thus not just about the physical effects of development such as pollution or sprawl. It is also about the feelings, attitudes, and sensory experiences nurtured by the environment that contribute to civic consciousness and identity. Just as civic attitudes and the "habits of the heart" that Tocqueville saw as critical to the success of democratic communities affect the way in which physical space is developed, so too does the sense of place and experience of nature influence our civic sensibility and public consciousness.

The Vehicles of Environmental Change

What, then, are examples of this dialectic in action? How is the relationship between community and environment played out on the stage of the American landscape? Before we can answer these questions, we must first look at the ways in which civic conditions are translated into environmental effects. This occurs through what I define as the vehicles of environmental change: development, production, and consumption. These vehicles are the human forces that essentially mediate the relationship between communities and the environment, converting civic conditions into environmental effects and vice versa. They are the proximate, not ultimate, causes of environmental change; they are the conduit through which civic consciousness becomes physical reality.

Development

Development refers to the public and private actions that affect the way in which land and natural resources are used for the purposes of production, consumption, and conservation. To some extent, all environmental effects, whether air and water pollution, contaminated land ("brownfields"), or habitat destruction, are a function of development, which can also be called land use. By definition, the use or development (as opposed to conservation) of land and natural resources entails environmental impacts. The clearing of forests, filling of wetlands, paving of open space, and other similar environmental impacts are some of the more adverse consequences of development. Suburbanization, the most significant land use trend of the past half-century, has caused massive air pollution as a result of the ever-increasing vehicle miles traveled by commuters, as well as destruction of habitats owing to the development of subdivisions, office parks, and shopping malls. Indeed, the very institution of environmental law and policy, designed to control and mitigate pollution, can be understood as a response to the land use and development decisions of the 1950s, when hordes of upwardly mobile Americans left the cities for the clean air and green lawns of the suburbs only to realize that they brought the smog, traffic, and sprawl of the cities in their wake.[13]

Development and land use are a precursor to production and consumption in that they determine the arrangement of physical space and the distribution of human activities—residential, commercial, manufacturing, industrial, recreational—on specific parcels of land. Before land and natural resources are put to a specific use, thus engaging the forces of production and consumption, land use and development decisions determine where and to what extent natural resources can be used or extracted and for what purposes. Yet land use and development are themselves driven by the forces of production and consumption. For instance, producers develop land and natural resources for raw materials to be used in the production process and to construct the facilities that allow producers to make their products; similarly consumers purchase land to build housing, churches, or parks, and they patronize businesses whose facilities and products cause environmental impacts such as hazardous waste or polluted run-off.

Development, production, and consumption, as well as their environmental effects, are thus overlapping and reciprocating forces, aligned in a dialectical fashion. However, while production and consumption influence development and land use decisions (a lumber mill seeks to harvest timber on a tract of privately owned old-growth forest; a housing shortage prompts a municipality to consider subdividing ten acres of city-owned land), these decisions determine whether and to what extent production and consumption will be allowed in the first place.

A Brief History of Development and Land Use in the United States Prior to their formal regulation, first through public health regulations in the late nineteenth century and later through zoning laws in the early twentieth, development and land use in the United States were driven solely by the forces of conquest, expansion, and immigration. The constant and deliberate westward settlement of vast territories formerly occupied by Native American tribes or controlled by European sovereigns and the concomitant exploitation of natural resources largely define the first one hundred years of American history. The mid-nineteenth century, for example, came to be known as the free land era because the U.S. government gave land away to homesteaders to encourage settlement of remote areas, thereby relieving the eastern cities of pressure from overcrowding and fulfilling Jefferson's dream of a nation populated by yeoman farmers. This was the age of the railroad, the gold rush, and manifest destiny. The widespread settlement and development of land was also celebrated as the visible confirmation of

American society's progress.[14] From the time of the first English settlers in the seventeenth century, Americans believed that land development was a religious imperative, inspired by a sense of divine providence.

By the 1890s, the Pacific coast had been fully settled, and much of the East was urbanized. New England, which was 90 percent agricultural land in 1810, was the first region to become industrialized, around 1850.[15] Boston, Lowell, Worcester, New Haven, Hartford, Providence: these New England cities were among the world's leading industrial centers by the middle of the nineteenth century. The nineteenth-century city, architectural historian Witold Rybczynski reminds us, was "an industrial city, and industry . . . was extremely dirty."[16] Coal-burning factories and power plants, stockyards, slaughterhouses, and horse-drawn vehicles marked the urban landscape and produced what Lewis Mumford called the "most degraded human environment the world had yet seen."[17] Urban environmental conditions forced cities like New York in the early 1900s to adopt land use controls—zoning rules—to protect residential districts from the pollution and waste generated by nearby industrial facilities.

Environmental degradation, coupled with explosive population growth brought about by immigration, led to large-scale suburbanization in the early twentieth century (suburbs had first been developed in the 1830s and 1840s by the landscape architect Andrew Jackson Downing). By the 1920s, suburbs were growing faster than cities; by the 1930s, cities like Boston were losing population for the first time.[18]

The same desire for land and prosperity that drove the westward expansion and urbanization of the 1800s was thus transformed into the suburban migration of the 1900s. Immediately following World War II, the first suburban subdivisions were developed, spurred by the construction of major roadways (including the interstate highway system in the 1950s) and the availability of federal mortgage subsidies. Without much in the way of landscaping or planning, the postwar suburbs were explicitly designed for cars and shopping centers. Developed in 1947, Levittown, on New York's Long Island, became the prototype of mass suburban development, with its seventeen thousand identical homes and countless miles of streets. Based on inexpensive housing and standardized design, the massive subdivision was made possible by automobile access. Personal mobility obviated the need for town centers, streetcars, and railroads. All the subdivision required was housing, streets, and a shopping center.

Over the past forty years, suburbs have attracted more than 75 million people. By 1990, more Americans lived in suburbs than in cities and rural areas combined.[19] Underlying this suburban boom has been the latest wave of land development, commonly referred to as sprawl—the literal spreading out of unplanned, uncoordinated development across the American landscape, in the process transforming ecosystems, open space, and woodlands into subdivisions, shopping malls, roadways, and office parks, each with associated environmental impacts.

Sprawl reveals itself in raw numbers. According to the Sierra Club, from 1970 to 1990, more than 19 million acres of rural land across the country were developed; 400,000 acres a year are lost to subdivisions, shopping malls, office parks, and roadways.[20] In the metropolitan Chicago area, developed land area increased by 46 percent between 1970 and 1990, though the population grew by only 4 percent. During the same period, suburban Los Angeles, whose population grew by 45 percent, saw an increase in developed land of 300 percent.[21] Oregon loses 2,000 acres of agricultural land per year to sprawl; Colorado loses 50,000 acres per year.[22] Scottsdale, Arizona, which in 1950 had a population of barely 2,000 people on one square mile, today is home to 165,000 residents with a land area three times the size of San Francisco; between 1993 and 1996, 25,000 acres of Sonoran Desert disappeared beneath the new roadways and parking lots of metropolitan Phoenix.[23] As the journalist Bill McKibben has observed,

> With each year, the ring of suburbs spreads a little farther out, the roads become more crowded, the margin for wildlife becomes slightly smaller. That endless growth places real stress on our supplies of everything from water to silence, from farmland to solitude. Such growth even strains our democracy. When the constitution was ratified, each member of Congress represented 30,000 voters; now it's 570,000. But there's a bigger problem still. Americans . . . contribute far more per capita to the world's environmental problems than anyone else.[24]

Recently, while some metropolitan areas like New York, Los Angeles, and San Francisco have lost population (at the same time dramatically increasing their geographic area), remote rural areas have experienced significant population growth and development. Reversing a decade-long decline in rural population, recent studies show that rural counties are booming, attracting thousands of new residents and swallowing up huge tracts of once-barren land. Lyon County, Nevada, for example, had a population of 15,600 in 1980. In 1996, that population grew to 31,500; by

2010, it is expected to be close to 50,000.[25] With the construction of each new house, street, and office building, rural land and natural resources are being lost, ushering in human activities that will create additional environmental impacts.

Meanwhile, just as the suburbs began spreading rapidly throughout the land fifty years ago, the cities witnessed major renewal projects intended to clean up the slums and industrial detritus that had accumulated over the preceding hundred years. From Boston and New York to Chicago and Los Angeles, urban renewal efforts transformed the spatial order of cities by erecting massive roadway systems and public housing developments, often literally on top of what were formerly neighborhoods and parks, while cutting off access to natural resources like waterfronts and woodlands.

Moreover, the mixed-use neighborhoods of the past, developed often organically over time, were replaced by single-use zones meant to separate uses and people. Urban renewal succeeded at making it easier for more affluent residents to leave the cities for the suburbs with new highway infrastructure, while making it harder for less mobile residents to live day to day.

Liberalism, Private Property, and Development Land use and development in the twentieth century have been so extensive it is believed that 80 percent of everything built in the United States was constructed within the past fifty years.[26] At the core of this unbridled growth is the tradition of Lockean liberalism. Since the time of the first settlers in the 1600s, American land use has been based on the English private property system, which encourages property owners to regard their land and its products as commodities for exchange and profit. Land has always been seen as a form of capital, a commodity consumed for the express purpose of creating wealth. Recall that private property rights are an inherent component of a capitalist system; they provide owners a reasonably secure expectation of continued ownership, allowing owners to invest in improvements to their property, and they provide buyers and sellers of property the knowledge they need to engage in exchange.

Notwithstanding a significant tradition of public parks, such as the majestic landscape projects of Frederic Law Olmsted, created as a response to the overdevelopment of America's cities and towns, land has rarely been considered a social resource or public trust. The City Beautiful movement of the early twentieth century, for example, which sought to incorporate civic and public values in the planning and design of American cities, failed largely because of opposition from commercial and private property interests.[27] Even the establishment of the national park system, whose origins date back to the founding of Yellowstone in 1872, was as much a concession to private property interests as it was to high-minded preservationists. The early parks were created, in large part, because their location was deemed too remote and rugged by the businessmen and legislators who would otherwise object in the interest of land speculation.[28]

"To build a better motor we tap the uttermost power of the human brain," the environmentalist Aldo Leopold despaired; "to build a better countryside, we throw dice."[29] American development patterns have tended to defy coherence or planning. With the exception of public parks and state and federal lands (which are not completely off limits to development and resource exploitation), the American land ethic has been thoroughly private in nature. Development is a powerful vehicle by which individuals and communities make their mark, sometimes indelibly. And because development activities, like architecture, manifest individual and community values in physical form, these markings enable us to get a glimpse of what civic life in America looks like.

Production and Consumption

Before we undertake to read the civic landscape, we must first examine the other vehicles of environmental change, production and consumption—what happens after development decisions have been made. Close companions to development, these forces too shape the environment and help reveal the nature of American civic life.

Production Production refers to the material practices of economic production: the extraction, processing, and exchange of resources. In addition to environmental effects related to land use, such as clear-cutting and mining, production entails energy and materials flows, which generate environmental impacts and wastes. At every stage in the life cycle of a product, from resource extraction to finished product to consumption, environmental impacts and waste are produced.

Consider the production of automobiles, a vast and complicated industrial process. Viewed at its most basic level, this process extracts ore and other raw materials, which are then manufactured into the car's material components, which are then assembled into an automobile and sold on the open market. Throughout the entire life cycle of the product, environmental effects are generated. The mining of raw materials causes significant environmental impacts, from air pollution to water and soil degradation. By its very nature, mining is an extremely destructive practice; considerable impacts are thus part and parcel of the process. Mining also requires energy, such as oil, to power the machines used to extract the ore. That energy is itself extracted and processed, which entail environmental effects—oil spills from tankers and trucks, for example, or destruction of habitat due to oil exploration and development.

Transporting raw materials to the manufacturing site is another stage in the production process requiring energy and causing environmental harm, such as air pollution from trains and trucks. The actual manufacturing process is energy intensive and involves large-scale emissions of pollutants from multiple sources within the plant, which affect all environmental media. From easily identifiable sources at the facility like smokestacks and outflow pipes, to less discernible pollution sources like the trucks that service the plant and emit toxic exhaust or the parking lots employees use, where oil-contaminated runoff from cars leaches into soils, groundwater, and streams, the traditional manufacturing plant generates environmental impacts from multiple sites and in multiple ways.

That is not all. The materials used in manufacturing automobiles, from metals to plastics, are often made of hazardous substances that expose workers, plant neighbors, and consumers to health risks at various stages in the product's life cycle. Moreover, those materials are themselves usually extracted and produced at other sites and are then transported to the automobile plant. In other words, the production process for a single product such as a car actually amounts to a series of many overlapping processes involving that product's components, the components' components, and so on.

Eventually the car is produced and delivered to market for sale to the consumer. The product delivery stage, like every other stage in the product's life cycle, requires energy for transportation and storage and, consequently, produces waste like air and water pollution from trucks and freight ships.

Production thus entails a vast complex of materials and energy flows, which result in wastes and pollution. These are the environmental effects that mostly occur after the initial impacts from development are generated, although development and production impacts can be one and the same. Filling wetlands, cutting down trees, and paving open space, for instance, are both land use and production impacts, causing ongoing environmental harms such as diminished water filtration and carrying capacity, and soil erosion and contaminated runoff, respectively. These effects continue along with the secondary effects caused by production activities.

Thus, each year in the United States nearly 800,000 acres of trees are cut down (for land use and production), hundreds of billions of pounds of chemicals are used, and 250 trillion pounds of waste (including wastewater) are generated as a result of production activities. American companies waste 300 billion pounds of organic and inorganic chemicals used for manufacturing and processing and generate 700 billion pounds of hazardous waste caused by chemical production. Thirty-four trillion pounds of waste per year are produced as a result of extracting gas, coal, oil, and minerals. For every 100 pounds of product we manufacture, we create at least 3,200 pounds of waste.[30] These figures are just a sample of the environmental effects of production.

Despite the multitude of environmental impacts that accompany traditional production activities, the cost of those impacts has typically not been accounted for by producers, whether in their balance sheets or in the price that consumers pay for their products. Historically, environmental impacts have been treated as a necessary and inevitable part of land use and production; their costs have thus been assumed away, neither borne by producers nor passed on to consumers. Economists describe this phenomenon as the externalization of environmental costs. Externalities, like waste and pollution, are the stranded costs of production; they simply float about, orphans of the economic system. Conversely, by internalizing these costs, producers are able to assess accurately the true cost of production in terms of environmental impacts and thus take appropriate steps to avoid or minimize those costs. The externalization of costs has meant that until the introduction of environmental regulations in the early 1970s, pollution and environmental damage were written off as an essential fact of modern, industrial life, with virtually no one held accountable.

Our environmental law and policy system has had some success in dealing with the environmental problems created by production. But that success has been limited, owing in large part to the lack of the civic resources necessary to effect lasting environmental protection results.

Consumption Consumption follows and feeds production. Consumers purchase land and other natural resources in their undeveloped state, and they buy the products that producers sell. Consumer demand dictates what and how much producers produce, thus influencing the environmental effects of production. For example, consumer demand for bigger, heavier, gas-guzzling automobiles, such as sport utility vehicles, increases reliance on petroleum products, whose production entails significant environmental damage; consumer appetites for foodstuffs and fabrics whose production relies on the use of pesticides promote the contamination of soils and water from runoff, in addition to health risks to workers and consumers themselves. As well, consumer demand for discount goods from high-volume retail outlets has fueled the spread of large, "big-box" superstores (often exceeding 200,000 square feet and three stories) and shopping malls, whose footprint alone chews up countless acres of raw land and wetlands and whose parking capacity and location encourage significant amounts of automobile traffic, thus causing air pollution.

The decisions and actions of consumers also directly generate environmental impacts. Driving cars and trucks, especially big, heavy vehicles (including so-called light trucks), produces air pollution and oil-contaminated runoff from roads and parking lots; purchasing houses in large-lot, suburban subdivisions contributes to the loss of habitat, while exacerbating air quality impacts because these subdivisions encourage dependence on automobile transportation; using products made of virgin materials requires the destruction of forests and other ecosystems.

According to Paul Hawken, Americans have the largest material requirements in the world, consuming on average 125 pounds of material per person every day. This consumption includes everything from fuels, mined materials, and industrial minerals and metals to forestry and agricultural products. Each year Americans waste more than 1 million pounds of material per person, including 3.5 billion pounds of carpet and 6 billion pounds of polystyrene sent to landfills, and 25 billion pounds of carbon dioxide released in the air. Since 1900, Americans have disposed of well over 650 million vehicles.[31]

Consumption produces pollution and waste and eats up land and natural resources. For over three hundred years, Americans have consumed nature at unprecedented rates. Such consumption, fueled by the forces of the market and mass production, has over time eroded Americans' connection to nature and sense of reciprocity. By consuming more and more, it seems, we are left with fewer places and ecosystems with which to connect.

Ultimately, development, production, and consumption shape the environment and the physical conditions in which human communities live, work, and play; they determine the quality of our lives and our very survivability as a species. They are also the means by which our everyday actions become physical marks on the American landscape, either temporary or permanent.

The Environmental Conditions of American Democracy

Land use, production, and consumption are the proximate causes of environmental change—the vehicles by which communities shape their environment. In turn, environmental conditions influence the social conditions of communities in a variety of ways. Environmental impacts fall under several categories according to environmental media, such as air and water. Impacts like air pollution or filling of wetlands straddle and traverse media-based categories because they, like environmental media, are not stagnant or transient. Rather, they tend to be dynamic and durable. Thus, for instance, a leaking underground storage tank filled with oil might result not only in contaminated land, or brownfields, but also contamination of a local water supply or wetland. The clear-cutting of forests causes both water and air pollution in that clear-cutting results in the loss of erosion control and diminished air filtration, respectively. Emissions from smokestacks and automobiles create smog and ozone as well as acid rain, which eventually falls on trees and in lakes and streams, damaging those resources.

The ways in which pollution is controlled or mitigated often result in merely displacing pollution from one medium to another. So, for example, when air pollution is controlled through the use of smokestack scrubbers, the amount of air emissions is reduced, but a hazardous filtrate is produced that must be disposed of in a landfill; the contamination has been displaced from the air to the soil. Similarly, when contaminated sediment from lakes or rivers is dredged, it must either be disposed of in a landfill or incinerated (leaving behind a toxic ash that must be shipped to a landfill for disposal), thus resulting in the transfer of the pollution from water to land or air, or both.

Underlying all environmental impacts is human health and the survival of the human species. Americans care about environmental protection not simply because we are concerned about preserving beautiful wilderness areas or saving endangered species. Ultimately we are motivated by our interest in self-preservation. Lung-burning smog, mercury-laden fish, soil contaminated with toxic materials, development of open space: these environmental impacts imperil our lives and undermine our quality of life. The fundamental fact that human health is totally dependent on the health of ecosystems compels corrective action.

Environmental Indicators: America's Physical Health

As the environmental ethic holds, all things are connected and must be treated accordingly, especially when designing environmental protection strategies. Yet for the purposes of viewing a part of the vast landscape of environmental issues, we will examine environmental conditions according to a set of three basic categories, mindful of the interrelatedness of those conditions and the human health issues that underlie them: toxics and brownfields, water pollution and wetlands degradation, and indoor and ambient air pollution. (Other categories, such as forests, wildlife, and endangered species, are equally important but are omitted in the interest of brevity.)

The discussion that follows is intended to paint with broad brushstrokes a picture of current environmental conditions, without much in the way of explication of the causes of those conditions. Development, production, and consumption are the vehicles of environmental harm, but they are not the ultimate causes. Each derives from discrete human decisions and actions that are themselves determined by larger social forces that influence and constrain those decisions and actions.

Toxics and Brownfields For thousands of years, human societies have used and developed poisonous substances in producing the materials and products necessary to sustain and improve living conditions. In ancient Rome, for example, engineers used lead, a naturally occurring element that happens to be extremely toxic, affecting humans' neurological functions, for the sewer infrastructure required by the imperial city's burgeoning population. Lead is extremely durable, making it attractive for pipes and other materials designed to last a long time. (The modern word *plumbing* derives from the Latin word for lead, *plumbum.*) Other hazardous substances found in nature, like mercury, copper, cadmium, chlorine, and arsenic, have similarly been used over time for a variety of production purposes, from manufacturing paper products to fabricating metals to farming and agriculture.

In modern times, new, synthetic compounds, many of them as toxic as they are commonplace, have been created to make life easier and cheaper for producers and consumers alike. Developed in the nineteenth century, refined petroleum, used as fuel to power everything from gas lamps and furnaces to automobiles and electricity plants, is hazardous in its liquid state, containing hydrocarbons and other toxic chemicals, and when combusted, releasing a variety of waste gases and particles that contribute to global climate change and smog. Although it no longer contains lead, the gasoline used in cars in the United States is composed of a mixture of over one hundred organic compounds, many of them highly toxic, including benzene, known to cause severe blood disorders and leukemia. Petroleum is also an essential ingredient in plastics, which, when burned, produce carcinogenic compounds like dioxin, the most toxic chemical known to science. Asbestos, a processed mineral commonly used for insulation in building materials, contains cancer-causing chemicals that can get lodged in human lung tissue when asbestos becomes airborne or friable. DDT, the insecticide made notorious by Rachel Carson's *Silent Spring* in 1964 and subsequently banned in 1972, weakened the eggshells of birds of prey like osprey and bald eagles that consumed fish poisoned by DDT-contaminated runoff, resulting in their near extinction. Chlorinated compounds such as polychlorinated biphenyls (PCBs), a heat-resistant material used to insulate electrical transformers, are known to cause cancer, skin disease, and reproductive disorders; when they are combusted, they produce deadly dioxin. Lethal atomic materials are also a product of the modern age. Spent uranium and plutonium, for instance, are both extremely hazardous to human health, and persist in the environment for tens of thousands of years.

Natural and manufactured hazardous materials abound in the postindustrial era, as both substances used in the normal course of industrial activity and the waste or by-product of that activity. A host of environmental laws exists to regulate the manufacture, use, handling, and disposal of hazardous substances and wastes. Yet, with few exceptions, those laws do not *prevent* the use of toxic materials or generation of hazardous waste; they merely purport to control their environmental and health effects.

Under the federal Emergency Planning and Community Right-to-Know Act (EPCRA), for example, all manufacturers and businesses with ten or more full-time employees are required to report the locations and quantities of hazardous chemicals if the facility manufactures or processes over 25,000 pounds of the approximately 300 designated chemicals or uses more than 10,000 pounds of any designated chemical. Every year, over 80,000 reports are submitted to the Environmental Protection Agency (EPA) and state governments by more than 20,000 facilities.[32] EPCRA thus serves to assist communities in identifying where and how much toxic materials are being processed or used. It is an information-generating tool. It does not, however, prohibit hazardous materials from being processed or used.

Almost everything we use, eat, or see during the average day is composed of, packaged in, or treated with some toxic material. Most buildings constructed before the mid-1970s are coated with lead paint or contain asbestos insulation. Most paper products are bathed in chlorine to bleach them. Almost every car, bus, truck, and airplane burns petroleum. Nonorganic fruits, vegetables, cotton, and other agricultural products are sprayed with chemical pesticides to ensure a high yield. In the United States alone, 860 active-ingredient pesticide chemicals are formulated into 21,000 commercial products registered with the EPA. Of those, 278 are directly applied to raw agricultural crops.[33]

Paradoxically, toxic materials are the essential ingredients of the good life in the modern era. Petroleum, plastics, pesticides, and other toxics allow us to move around faster, eat more, and stay warmer or cooler than we would without them. Moreover, they reduce the cost of producing and purchasing goods, enabling more Americans to enjoy a standard of living unavailable to earlier generations. Indeed, toxic materials and the technologies that have accompanied them have made it possible for people who live in industrialized nations to live longer and more comfortably than ever before. This is perhaps the greatest paradox of them all: industrialized nations are better able to feed, shelter, and care for their citizens than are less developed societies, as evidenced by the longevity of their citizens, rates of infant mortality and morbidity, rates of poverty, and the like.

Yet when we look beyond average life expectancy and other similar statistics, we see in industrial societies like the United States health risks and environmental damage that invariably stem from exposure to and releases of great quantities of hazardous substances. Thus, in yet another paradox, it is likely that increased incidences of certain diseases like cancer, bronchitis, and asthma are directly related to the prolonged exposure to environmental hazards that a longer life span and greater material prosperity allow.

Moreover, despite undeniable progress in the United States in the past several decades in addressing environmental harms through law, technology, and market reforms, a multitude of hazards persist. For example, occupational exposure to toxics kills approximately 60,000 American workers per year.[34] Pollution from PCBs and other heavy metals has left over 40 percent of America's rivers, lakes, and streams unsafe for swimming and fishing.[35] In New England alone, 40 percent of the coastal estuaries have elevated levels of sediment contamination from toxic releases.[36] Though DDT is gone, lead and mercury still endanger fish-eating birds like loons. Close to 2 million young children, or about 8.9 percent of preschoolers in the United States, have lead levels in their blood that are considered dangerous.[37]

Spills and releases of hazardous substances have resulted in countless thousands of acres of contaminated land, or brownfields. Over the past several decades, as industrial and manufacturing activity has declined across the United States, and especially in older urban centers, due to demographic changes, national and global competition, and the recent shift from a manufacturing-based economy to a service-centered, knowledge-based one, thousands upon thousands of abandoned, contaminated sites have been left behind.[38] The EPA defines brownfields as "abandoned, idled, or under-used industrial and commercial facilities where expansion or redevelopment is complicated by real or perceived environmental contamination." Brownfields come in all shapes and sizes, ranging from abandoned incinerators to closed gasoline stations to former plastics manufacturing facilities.

Estimates of the number of brownfield sites in the United States vary. The U.S. General Accounting Office has estimated that there are between 130,000 and 450,000 contaminated commercial and industrial sites around the country. In New York City alone, there are approximately 6,500 tracts of contaminated land, amounting to more than 3,300 acres of brownfields.[39] Current estimates place the cost of cleaning up the nation's brownfields at $650 billion.[40] The sites vary in size from less than 1 acre to hundreds of acres. Based on some estimates, the total size of all brownfields, if assembled in one place, would cover Los Angeles County.

Brownfield sites can be found all over the country, in urban, suburban, and even some rural communities. Many sites are located in the Northeast and Midwest where, historically, much of the economic activity was industrial in nature. Although both cities and suburbs suffered the effects of industrial flight and economic transformation, suburbs have generally been able to attract reinvestment capital by expanding other employment sectors. Their inner-city counterparts have not been so fortunate. As a result, brownfields tend to be disproportionately concentrated in distressed urban areas, that is, in people of color and lower-income communities.

By their very nature, brownfields pose a public health risk and an environmental hazard. Until it is properly and permanently remediated, contamination from these sites can spread to neighboring properties, affecting the soil and groundwater. More important, abandoned sites are frequently left open and unsecured, creating an attractive nuisance. Even fences and warning signs (often sparingly posted and usually only in English) fail to keep people out. They are frequently located in the midst of residential neighborhoods and retail centers; children from time to time wander onto these sites and homeless people use them as temporary shelter. Limited only by the level and type of contamination, abandoned, contaminated sites affect people's health directly daily.

The adverse effects of brownfields go beyond their immediate environmental impact. Vacant lots, contaminated or not, attract illegal dumping activities. Many, if not most, of these vacant lots are used as illegal solid waste disposal sites. These dumping activities bring along their own environmental threats, ranging from groundwater pollution to rodent infestations. In addition to brownfields, toxic waste sites are prolific in the United States. According to EPA administrator Carole Browner, one in four

Americans lives near a hazardous waste dump. Further, several studies have shown that active commercial hazardous waste facilities are twice as likely to be located in communities of color than in white communities and that income levels and home values in the 369 communities with these waste facilities were significantly lower than in surrounding counties.[41] More than 26 million people of color live in communities with inactive hazardous waste sites, which number greater than 18,000. This is slightly greater than the percentage of white people who live near such sites.

Other kinds of industrial hazards also find their way disproportionately into communities of color. Of the counties in the United States with the highest national rankings for twenty-four measures of industrial hazards, including number of smokestacks, tons of hazardous chemicals released into the air and water, hazardous waste facilities, and occupational exposures, people of color were 60 percent more likely than whites to live in the counties that ranked among the top 2 percent.[42] Latinos were 32 percent more likely to live there, African Americans were 85 percent more likely, and Asian Americans were nearly three times more likely than whites.

Toxics, hazardous waste, and brownfields are thus pervasive environmental threats in the United States. In every American city and town and across the countryside, these impacts exist in protean form: from blighted land, to rivers and lakes closed to recreation, to workers and children suffering from environmental poisoning. Disturbingly, these impacts tend to fall most heavily on people of color and lower-income communities. Although we no longer live in a society characterized by the kind of industrial filth and pollution Lewis Mumford warned about earlier in the twentieth century, we are still living within the environmental paradox of the industrial order: what environmental hazards give, they also take away.

Water Pollution and Wetlands Degradation The release of toxic materials into the nation's waterways is a significant contributor to water pollution and wetlands degradation. Just as producers and consumers are allowed to use hazardous substances, regulated by a complex system of environmental laws, so too are they permitted to discharge those substances as waste into lakes, rivers, and streams, also within a regulatory framework. Toxics thus lead not only to brownfields but to brownstreams as well. Remember that pollution defies media categories because it migrates, pushed along by its own physical properties and by the medium—air, water, or soil—into which it is released.

As a result of contaminated effluent from factories and sewage treatment plants, as well as polluted runoff from roads and agricultural land, 40 percent of America's rivers, lakes, and streams are too polluted for swimming or fishing. Large-scale industrial agriculture is the largest polluter, accounting for 70 percent of waterway pollution and fouling more than 173,000 miles of waterways with chemicals, animal waste runoff, and erosion.[43] Waste-saturated runoff from commercial hog and chicken farms is the most significant agricultural threat to America's water resources. A single farm can produce as much polluted runoff as a medium-sized city. In addition, as much as 180 million gallons of oil is dumped into rivers and streams each year by people who carelessly change their own motor oil; about 250 million gallons a year leaks from cars.[44] Further, 21 percent of the nation's watersheds have serious water quality problems, ranging from heavy metals contamination to eutrophication from sewage and agricultural sources (eutrophication refers to the depletion of dissolved oxygen in water resulting from the decomposition of algae and aquatic plants; it leads to algal blooms, high levels of turbidity, fish kills, and loss of benthic, or bottom-dwelling, animals), to siltation and low water flow. Thirty-six percent of American watersheds have moderate water quality; only 16 percent enjoy good water quality (EPA does not yet have enough information to assess the water quality for the remaining 27 percent of wetlands).[45]

Polluted water affects all Americans, but especially those who regularly consume fish caught in local rivers and streams for subsistence. Lower-income African Americans, for instance, in cities such as Detroit and Washington, D.C., routinely eat fish caught in waters that are highly contaminated, despite public health advisories. Native Americans, too, continue to fish ancestral grounds for subsistence, notwithstanding the routine fish kills, oil spills, and other hazards affecting those resources.

Further, more than 85 percent of inland waterways in the United States are artificially controlled by dams and other engineered structures, and at least half of the country's original wetlands (excluding those in Alaska) have been drained or filled.[46] In the two hundred years between the 1780s and the 1980s, more than 60 acres of wetlands were lost every hour.[47] As a result of the construction of dams and massive loss of wetlands to development, critical habitat for fish and wildlife has been lost.

In the West, where over the past century government-subsidized irrigation projects miraculously turned arid grasslands and desert into farmland from the Colorado River to California's Central Valley, only now are states beginning to reckon with the tremendous damage these projects wrought upon wetlands, wildlife, and the Native American tribes that traditionally relied on these natural resources. The Truckee River in northern Nevada, for example, site of the nation's first major federal irrigation project in 1902, was diverted for agricultural purposes at great cost to the environment and the Paiute tribe, which for thousands of years depended on the Truckee for subsistence fishing, water, and spiritual purposes. The Truckee dried up, becoming an open sewer and killing off a species of American cutthroat trout and the cui-ui, a Pleistocene-era fish that was the foundation of the Paiutes' diet. The project also drained Winnemucca Lake, a sacred marsh that was home to dozens of species of fish and waterfowl. The Truckee irrigation project was the forerunner to the great dams of later decades such as the Hoover and Grand Coulee, in Washington and Glen Canyon in Arizona. As a result of these projects, the epic salmon runs on the Columbia River were lost, and the magnificent Glen Canyon was submerged. Recently California, Nevada, and other western states have begun the process of revising their water policies to reclaim the wetlands, fish, and wildlife lost to the great irrigation projects earlier in the century in hopes of restoring the environment while serving the long-neglected water needs of tribes, cities and other areas deprived by agricultural irrigation.[48] Some, including legendary environmentalist David Brower, are even talking about draining Glen Canyon to restore it to its original natural splendor. In Wyoming, the Army Corps of Engineers and local municipalities are planning to remove levees from the Snake River built in the 1960s, allowing the river to reclaim its original floodplain and, ultimately, its ecology.[49]

The continuous destruction of wetlands has resulted in the loss of the important habitat and water filtration functions wetlands provide. In turn, this destruction has undermined countless ecosystems. In Boston, entire waterways have been filled and abandoned, though their aquatic names remain. The Back Bay, South Bay, and Alewife are the familiar names of places that were once thriving wetlands, having been filled and developed decades ago. No bays or alewives (herring) exist in these places anymore, just buildings, roads, and parking lots. Not surprisingly, they have lost almost all of their wetlands attributes. The Charles River, one of Massachusetts' most significant waterways, has not been safe for swimming or fishing since the 1950s. The Muddy River, a major Charles River artery flowing south to Roxbury and Brookline, is barely visible under the roadways, bridges, and buildings that comprise (ironically) the Back Bay. Just to the south, in New Bedford, the Acushnet River, site of the most famous eighteenth-century whaling port, contains one of the most polluted harbors in the world.

Moreover, development of floodplains and coastal areas for housing and other uses and the spread of impervious surfaces like roads and parking lots have resulted in annual multibillion dollar "natural disasters," in which rivers and seas crest and surge, temporarily reclaiming the land they have always called their own and leaving behind soggy, distraught homeowners who must question whether living so close to water is worth it. "Major floods used to be quarter-century events in the Puget Sound region," the journalist Timothy Egan reports, "Now, almost every winter, the river valleys east of Seattle swell with coffee-colored water and overwhelm roads and farms."[50] Recall the great Mississippi River flood of the 1993 and the frequent coastal mudslides in southern California.

Meanwhile, three centuries' worth of polluted effluent and runoff into America's rivers, streams, lakes, and harbors has rendered a large portion of them lifeless and murky. Though no longer as prone to bursting into flames as Cleveland's Cuyahoga River was in the early 1970s, America's waterways have a long way to go on the road to recovery. In how many urban areas can residents swim or fish in their local waterways? How often do we hear about an oil spill or raw sewage release into our rivers and streams? How many native tribes must contend with annual fish kills in their ancestral fishing grounds? The unsettling answers to these questions provide a glimpse of the general condition of America's water resources.

Ambient and Indoor Air Pollution According to a 1997 *New York Times* poll, air pollution is the environmental problem mentioned most often by Americans.[51] *Global warming, ozone depletion,* and *Los Angeles smog* are part of the familiar lexicon of environmental problems. Notwithstanding major recent gains in cleaning up the nation's ambient air, from the Los Angeles Basin (where for the first time in over fifty years the majestic San Gabriel Mountains, just to the north, are frequently visible from the downtown area) to the rust belt, one in three Americans still lives in an area where the air quality does not meet federal health standards. Generally Americans' cars burn gas more efficiently and our factories spew fewer pollutants into the atmosphere, yet air pollution remains a grave problem in the United States.

America's cities are home to some of the worst ambient air pollution. Despite its recent successes, Los Angeles's air pollution levels are still the worst in the nation, exceeding federal standards for carbon monoxide and ozone in 1997 on fifteen days and sixty-eight days, respectively. Houston and New York are not far behind. Vehicles, construction equipment, manufacturing and industrial plants, and commercial activities such as dry cleaners and retail gas stations are the major sources of urban air pollution. Whereas most people tend to associate air pollution with large industrial facilities and tall smokestacks, most air pollution, especially in urban areas, comes from a variety of small sources, like cars, dry cleaners, and even wood-burning stoves. Nitrogen oxides, volatile organic compounds, toxic particles, and carbon monoxide are among the most common urban air pollutants. Such pollutants lead to the formation of ground-level ozone and the smog and haze that burn the eyes and contribute to asthma, the leading cause of chronic illness among children and of school absenteeism due to sickness. In the past decade, the number of Americans with asthma has increased by 42 percent and continues to rise, elevating asthma to the status of the "environmental disease of the decade," according to EPA. In Boston alone, it is estimated that more than 100,000 children suffer from asthma. Children from lower-income neighborhoods and neighborhoods of color are most vulnerable. A study by the Centers for Disease Control and Prevention found that between 1980 and 1993, young African Americans were four to six times more likely to die of asthma than young whites.[52]

People of color in general are the most affected by urban air pollution. They live in cities to a far greater extent than whites; close to 60 percent of all African Americans and approximately half of the Hispanic and Asian populations make their homes in America's central cities, while only a quarter of the white population does. People of color live in disproportionate numbers next to polluting facilities and congested roadways, where vehicles emit thousands of tons of air pollutants each day. A recent study found that in California, people of color are nearly three times as likely to breathe air polluted by fine particles as whites. The study found that 54 percent of the air monitors in communities of color had fine-particle readings above the 1997 EPA standard, while only 19 percent of the air monitors in white communities had readings above the standard.[53]

With greater than 1.4 billion tons per year of carbon emissions, the United States is consistently the world's largest contributor of so-called greenhouse gases, the emissions from the burning of fossil fuels that cause global climate change. The United States also has the highest per capita carbon emissions, 5.25 tons annually, which is more than seven times that of China and twenty-five times that of India.[54]

A principal factor in the United States' role as the leading global climate change culprit is Americans' love of big, heavy automobiles. Transportation sources account for approximately one-third of the greenhouse gases emitted around the globe. In the United States, cars and light trucks comprise one-fifth of the emissions of these gases. Light trucks, which include the now-ubiquitous sport utility vehicle, make up roughly half of the cars on the road, up from one-sixth in the 1970s. With drastically higher levels of emissions, lower fuel economy, and less stringent emissions standards than cars, the popularity of light trucks is forcing EPA to raise its estimates of nitrogen oxide and carbon dioxide emissions from family vehicles in 2020 significantly.[55] Almost single-handedly, light trucks are undermining the nation's effort to achieve clean air goals.

Not only is the explosion of light trucks on America's roads exacerbating the nation's air quality problems, not to mention the world's, but so too is the ever increasing vehicle miles traveled (VMT) by Americans. In 1995, for example, cars and light trucks were driven 2.2 trillion miles in the United States. This is nearly one-tenth of the distance to the nearest star outside our solar system. Between 1970 and 1995, passenger travel nearly doubled in the United States, growing by an average of 2.7 percent a year. Passenger miles per person increased during this period from 11,400 miles to 17,200 miles.[56] In California, where the population grew by 50 percent from 1970 to 1990, VMT grew by twice as much. Just to the north, in Seattle, Washington, VMT quadrupled from 1980 to 1990, while the population rose by only 22 percent.[57] The unpopularity and inadequacy of most mass transit systems, coupled with the great commuting distances in many suburbs and rural areas, fuels the problem. With every increase in VMT, air quality in the United States suffers.

Older power plants that burn coal and oil are also a major contributor to air pollution. Such plants, grandfathered into U.S. air pollution laws, emit huge amounts of nitrogen oxides, carbon monoxide, sulfur dioxide, particulates, and other pollutants with virtually no pollution controls. They are responsible for much of the acid rain that falls on the eastern states and for the ozone smog that travels from the Midwest to the Northeast.

Although indoor air pollution is typically ignored as a major environmental threat, it is as dangerous as ozone smog and global climate disruption. The neglected twin of ambient air pollution, indoor air pollution is an epidemic in the United States. According to EPA studies of human exposure to all air pollutants, indoor levels of pollutants may be two to five times, and sometimes more than one hundred times, higher than outdoor levels. These figures are especially disturbing because most people spend about 90 percent of their time indoors. Indoor air pollution is considered among the top five environmental health risks by EPA. Indoor air pollutants include mold and mildew from animal dander, decomposing biological matter such as cockroaches, and leaking roofs; radon; formaldehyde; and toxic gases from carpets and other furniture and building materials. Children are particularly vulnerable, especially those who live in substandard housing and attend poorly ventilated and maintained public schools. Many children in America's inner-city neighborhoods are exposed twenty-four hours a day, seven days a week to harmful indoor air pollution, contributing to asthma and other adverse health effects.

Indoor air pollution is the result of a variety of factors: the construction of tightly sealed buildings with reduced ventilation to save energy (recycled air); use of synthetic building materials and furnishings that emit hazardous gases; and use of chemically formulated personal care products (including chemicals used by most dry cleaners), pesticides, and housekeeping supplies. Four basic factors affect indoor air quality: sources of indoor air pollutants such as moldy, leaking roofs and ceilings; heating, ventilation, and air-conditioning (HVAC) systems; pollutant pathways; and the sensitivity of occupants.

A 1995 U.S. General Accounting Office study found that of 10,000 schools nationwide, half had poor indoor air quality conditions. A 1996 follow-up study reported that over 25 million children attend schools with serious indoor air quality problems. Of the approximately 4.5 million commercial structures in the United States, over 30 percent can be considered "sick" buildings because of indoor air pollution. The Occupational Safety and Health Administration estimates that it would cost $8.5 million to remedy the indoor air quality problems in America's commercial buildings.

Air pollution, whether indoor or ambient, thus persists as a significant threat to our environment and public health. Fueled by everything from VMT and power plants to deteriorating buildings and dry cleaners, air pollution remains as pervasive and insidious as ever, especially in the light of the hazards posed by global climate change, a recent scientific discovery. Although pollution-belching smokestacks no longer dominate the American landscape the way they did only a few decades ago, less easily discernible polluters now fill our skies and buildings with soot, smog, and mold.

Environmental Change and Environmental Quality

That environmental conditions in the United States continue to suffer under the pressure of the relentless forces of development, production, and consumption should hardly be surprising to even the most casual observer of the environment. The environmental indicators provide an accurate picture of the day-to-day realities of environmental harm that beset most Americans, especially those from lower-income and minority communities. As citizens, we witness this harm in a variety of ways: endless traffic jams that clog our roadways; lead that contaminates our backyards and poisons our children; the strip malls that blanket our suburbs and rural areas; the ozone smog that hovers above our cities and countryside; the lakes, rivers, and streams that are off-limits to swimming and fishing; and the asthma and upper-respiratory disease that afflict so many Americans, both young and old.

Yet as we learn from environmental historians like William Cronon, humans have been "manipulating ecosystems for as long as we have records of their passage."[58] Moreover, as Cronon declares, "living in history means leaving marks on the world; the question is, what kind of marks?"[59] At the beginning of the twenty-first century, it appears that many of the marks we have left on the American landscape are ones we should not be proud of: the brownfields, the polluted rivers and streams, the degraded habitats, the sprawling subdivisions and big-box superstores, and the acidic summer haze that dominate a significant part of the American environment. Especially in America's cities and older suburbs, environmental degradation is a defining characteristic. It is this physical reality that has prompted tens of thousands of Americans in recent years to flee these areas in search of the undeveloped landscapes of remote rural counties, continuing the tradition of rural flight and development that has shaped much of American history while swallowing up more land and habitat in the process.

The variety and scale of environmental impacts have driven a wedge between most Americans and their environment. Very few can claim a meaningful connection to the environment or sense of place because most Americans simply do not have an opportunity to experience nature where they live or work. As David Abram suggests, the forces of technology and development have robbed most of us of the chance to feel what it is like to live in the natural world, to experience the "oxygenating breath of the forest . . . the clutch of gravity and the tumbled magic of river rapids."[60] On account of poor land use planning, overdevelopment, environmentally unsound production methods, and unprecedented material consumption, the environment in which most Americans spend most of their time—the cities and suburbs—is at best uninspiring and at worst hazardous.

Nevertheless, the United States is not without its share of magnificent natural areas, still well protected and much loved by the American people. Perhaps more than any other country in the world, the United States attracts visitors who come expressly to see our national parks and wilderness areas, in many ways models of sound preservation and stewardship. In addition, many other areas of the country, from the pristine villages of New England to upscale midwestern suburbs to the cliff-hanging dwellings of the California coast, retain much of the natural beauty that characterized these areas prior to their settlement by Europeans and later immigrants. And yet, to the extent that environmental amenities such as clean air, beautiful landscapes, or swimmable streams are confined to certain discrete areas around the country, environmental quality as a whole is wanting. After all, the environment is not composed of separate, autonomous parts; it is a whole, seamless and connected. If environmental quality must be limited to a select few places, as in a small rural village or exclusive suburb, or is confined within the boundaries of a park or wilderness area, then environmental quality is inherently undemocratic and limited.

Just as modern institutions influence and socialize the behavior of individuals, creating an institutional mentality (the European sociologists Emile Durkheim and Max Weber first propounded this idea early in the twentieth century), so too does the physical environment. But like modern institutions, the modern American environment is very much the product of human choices and actions. With technology, industrialization, and the sheer force of numbers, we possess the power to shape and control our environment in a way no other society, let alone animal species, ever has.

Since time immemorial, environmental changes of enormous scale have occurred, many of them without any human influence. Simply consider the changes wrought by glaciation or volcanic activity, for instance. Humans as well have brought about significant environmental change. Before the arrival of European settlers in the seventeenth century, Native Americans made their mark on the American landscape through selective burning of woodlands to nourish soils and planting crops like corn for subsistence. As communities that tended to move from season to season and habitat to habitat and that maintained a deep spiritual connection to the natural environment, native tribes held their demands on the environment to a minimum. The definitive characteristic of Native Americans' relation to the environment was that it was essentially symbiotic and sustainable. Because the tribes were totally dependent on the availability of plants, animals, and other natural resources for subsistence and shelter, and espoused a belief system premised on the animistic idea of nonhuman nature as equal to humans, destruction of these resources was antithetical to their social and economic structure.

European settlers, however, with their fixed settlements, market-oriented economy, and private property regime, initiated the process of environmental change that remains to this day. By 1800, the New England environment had been completely transformed: beaver, deer, turkey, wolf, and bear had all but vanished; fences and livestock dominated the landscape;

soils were drier and exhausted; drainage patterns were less constant, and erosion and flooding were commonplace; local temperatures were erratic due to deforestation; and European pests and crop disease were prevalent.[61] This transformation amounted to widespread degradation of the environment and set in motion the land use, production, and consumption patterns that continue today. As a result, we essentially have two kinds of environment in the United States: the degraded and the protected.

The degraded environment is found where people live and work in large numbers—America's urban centers and suburbs, though rural areas are also subject to environmental harm. This is the transformed environment, where the only vestiges of dynamic ecosystems are the place names that refer to habitat or landscape features long since extinguished or developed. The protected environment, on the other hand, is located in designated parks and wilderness areas where very few people live and work. Vulnerable to pollution and overcrowding, though relatively pristine when compared to degraded places, parks and wilderness areas attract the greatest amount of attention and concern when it comes to environmental protection efforts. As the planner Robert Yaro has remarked, there are two land ethics in the United States: "Five percent of the nation was set aside as national parks, and treated as special places where we talked about things like 'America the Beautiful' and 'We, the People.' . . . But it seemed that different rules had to be obeyed throughout most of the rest of the land: either 'Take the Money and Run' or 'Private Property: Keep Out.'"[62]

Thus, ironically, Americans seem to care less about the places where they spend most of their time than the ones they visit occasionally, if at all. What accounts for this irony? What forces are responsible for the environmental degradation that so many Americans encounter and seem to abide day to day?

The Civic Causes of Environmental Decline

Driving the vehicles of environmental change are the civic conditions that influence and enable development, production, and consumption practices to occur in the first place. Development, production, and consumption activities do not take place in a vacuum. They are the product of human decision making and come out of a civic context, which, depending on the vigor and quality of civic life, constrains or directs those choices according to civic democratic principles. In other words, development, production, and consumption activities are proximate causes contingent on human decisions and actions, which in turn are shaped by a civic context.

To get to the root of the matter, we must examine the civic conditions in which environmental degradation occurs. Borrowing from the list of civic indicators discussed in chapter 1, we will look at social capital, political participation, racial and socioeconomic equality, and public investment and privatization and how each causally relates to environmental harm.

Social Capital

Social capital is the glue that holds communities together. It is the networks and associations in neighborhoods that allow individuals to realize public values, act on common interests, and solve public problems in concrete situations. As a public value and object of common interest, environmental protection is a highly valued currency of social capital. Community-based and grass-roots environmental organizations seeking to promote safe, healthy communities are a prime example of social capital. The more social capital a community possesses, the greater is its ability to solve problems and achieve positive environmental outcomes.

With the decline of social capital over the past several decades due, in large part, to the loss of the physical conditions that is—village-like settings, necessary to sustain social capital and the displacing effect of global capitalism—the networks that bind local environmental groups and other civic associations have deteriorated, creating a gap in communities' capacity to deal with public problems like environmental degradation. As the sociologist Richard Sennett claims, the working conditions of modern capitalism (downsizing, displacement) have eaten away at commitment and loyalty to place, to one's environment, among other objects of devotion.[63] Thus, environmental degradation tends to mirror the loss of social capital.

Acres upon acres of languishing brownfield sites, air pollution from the countless vehicle miles traveled by motorists sealed off from the rest of the world in their cars, and insidious unplanned, uncoordinated growth are environmental harms that speak of an impersonal, indifferent, and rootless society. They are assaults that deprive communities of a sense of place and experience of nature while transforming wildlife habitat, woodlands, and open space into cookie-cutter subdivisions, parking lots, and shopping malls—the raw materials of a placeless, market-driven society. Such transformation has resulted in places whose physical design is explicitly intended to keep people from each other and inside fortress-like structures that promote a sense of safety, not place. This is the basic architectural concept underlying most American malls built in the 1970s and 1980s,[64] and the impulse behind the proliferation of gated communities.

Moreover, in urban areas and older, first-ring suburbs, highway development, urban renewal projects and the spread of subdivisions have, Dolores Hayden states, "obscured large sections of the natural landscape and blotted out the cultural landscape of varied human activities."[65] Such devastation has robbed urban and suburban communities of the opportunity to develop the "casual public trust" Jane Jacobs describes as key to creating social capital.

These environmental effects are also disempowering to communities in that they inhibit people from coming together, congregating (the fundamental prerequisite of social capital), by creating dangerous and frightening physical barriers, such as brownfield sites or high-speed roadways, between community residents. By depriving citizens of the physical experiences that ground them to their communities, environmental degradation has further uprooted and displaced them, leaving them adrift and placeless. As Daniel Kemmis reminds us, the development of cooperation in a community, a precursor to the formation of social capital, depends on actual places, a physical environment that inspires public action. A sense of place, Kemmis explains, impels collaboration "because each . . . is attached to the place . . . [and is] brought into relationship with each other."[66] In this way, environmental harms reinforce the conditions of powerlessness and anomie that led to the degradation in the first place.

In urban and suburban communities alike, the fraying social fabric has created an environmental void being filled by the pervasive forces of global capitalism and the private property system. The result is the further exploitation of natural resources, environmentally unsound development, and the abandonment of polluted land and waters deemed too expensive to clean up. Without strong reserves of social capital, neighborhoods lack the capacity to assess, monitor, and prevent environmental harms. In effect, no one is minding the community.

Political Participation

Environmental degradation is also a function of political participation. Just as pollution and sprawl resulting from land use, production, and consumption undermine social capital, which leaves communities helpless to prevent environmental problems, so too does the lack of political participation result in a dialectic of civic and environmental degeneration. Without strong voter participation and related civic activity at the local, state, and federal levels, the decisions that determine such critical issues as the amount of investment in mass transit infrastructure, the stringency of environmental protection and public health laws, and the content of local zoning and planning ordinances are typically driven by special interest groups whose principal concern is usually avoiding regulation and the costs of achieving public goals like environmental protection. Weak political participation means simply that decision makers are not held accountable for their decisions, even when public opinion may be contrary to those decisions. Despite the vigilance of professional environmental organizations and other public interest groups, their efforts to ensure political accountability are greatly hindered by the lack of a strong, active constituency behind them.

It is no accident that the special interest groups that dominate environmental decision making tend to be associated with wealthy corporations whose profits derive from continuous, if not rapacious, economic growth, which is a function of development, production, and consumption, the same factors that lead to environmental harm. At all levels of government, corporations exercise disproportionate power over the political process because they underwrite political campaigns and operate in what is often a political vacuum, devoid of meaningful participation by ordinary Americans, which can counteract corporate influence.

As a result of corporate power and voter apathy, important environmental decisions are made regularly that permit and exacerbate environmental degradation. From state and federal regulations governing automobile emissions, use of hazardous materials, and cleanup of polluted land, to local land use decisions, maintenance of public parks, and enforcement of public health codes, environmental decisions necessarily take place within a public and political context. To the extent the public does not actively participate, a political void is created. Consequently, bigger, dirtier vehicles continue to roll out of Detroit's factories in huge numbers; new subdivisions and shopping centers are transforming huge tracts of remote rural land almost overnight; hundreds of thousands of abandoned and blighted brownfield sites, especially in lower-income and minority communities, dot the landscape; and industrial pollution, combined sewer overflow, and oil-tainted runoff from roadways are contaminating rivers and streams.

The recent trend in environmental law and policy toward market-based strategies, which we will examine in chapter 3, is further evidence of the success of corporations in influencing environmental decision making. These strategies refer to environmental policies that incorporate economic incentives to control pollution, such as tax credits, rather than penalties alone. Although not necessarily invidious or regressive, such strategies signal the ascendancy of corporate power over public policy. With little in the way of public involvement in environmental decision making, a privatized, corporate approach seems the only logical alternative to traditional command-and-control regulation.

As in the case of social capital, environmental degradation inhibits the sense of stake and common purpose that encourages and invites political participation. The feeling of powerlessness and futility that most voters express is bolstered by the physical environment in which they live, work, and play. As a recent *Boston Globe* story suggests, "Top-down [development] projects that laid waste to whole neighborhoods and threatened the stability of others" created a profound feeling of mistrust on the part of community residents, not only in Boston but across the country.[67] That mistrust has bred disaffection and disillusion.

Political participation and environmental accountability go hand in hand. Civic engagement is the force that tempers and moderates the activities of the private sphere, establishing a balance between the goals of public welfare and private gain. Especially at the community level, where voters and their elected officials are able to interact on an intimate and frequent basis and environmental impacts are most visible, a politically engaged citizenry can ensure that decisions affecting their environment and quality of life are made with their substantial input and review. In the absence of such engagement, environmental harm invariably follows.

Racial Equality

Environmental harms often follow the path of least resistance. As a result, disfranchised, economically disadvantaged communities of color tend to bear the greatest environmental burdens. Most American blacks, Latinos, and Asians make their home in the central cities, and especially inner-city neighborhoods, where environmental degradation is often at its worst. From toxic waste sites and water pollution, to ozone smog and dilapidated, hazardous buildings, to a dearth of parks and green space, America's inner cities present harsh conditions for humans and nature alike.

The ubiquitous tree of the inner city, the *Ailanthus altissimus*, or tree of heaven (introduced to the United States from China by Frederick Law Olmsted for Central Park), is a perfect symbol of that environment. At once imposing and elegant, with soot-streaked bark and exotic compound leaves, these trees can thrive where no other living things can grow: asphalt pavement, sewers, burned-out lots, junkyards, and concrete barricades, to name a few. Their emphasis is on survival, and survive they do. The ailanthus is nature's perfectly engineered answer to the challenge of the inner-city environment; it is the indicator species of urban America.

Minorities, including Native Americans, also face significant rural environmental threats. Because of the relative impoverishment of many rural minority communities, they are prime targets for lucrative offers from private waste companies and governments seeking to build landfills and incinerators for hazardous waste. As a result of what sociologist Robert Bullard calls "environmental blackmail," over two hundred indigenous communities have been propositioned by the waste disposal industry in recent years.[68] Further, with large reserves of such resources as uranium, copper, oil, and coal, tribal lands have experienced massive environmental degradation associated with the mining of these valuable resources by private companies. In tribal communities, as in other rural communities of color, toxic landfills, polluted rivers and streams, air pollution, and public health hazards are common.

Persistent racial inequality and prejudice in the United States have resulted in the physical segregation of communities, which has allowed undesirable land uses such as incinerators, dump sites, and other polluting facilities to be disproportionately sited in communities of color. The economic disinvestment and political disfranchisement produced by racial inequality have hindered the efforts of minority communities to mitigate and prevent such environmental hazards. For communities of color lacking a stable local economy and significant political clout, environmental harm is but one of a cluster of concerns. In fact, the same activities that often create the environmental hazards in communities of color also provide many of the jobs. Trash transfer stations, waste incinerators, auto body shops, metal fabricators, and other polluting facilities are typical inner-city businesses that offer significant employment opportunities for local residents. When it comes to choosing between a job and environmental degradation, the answer is regrettably clear, making environmental blackmail the common business practice in communities of color.

In effect, the inequality that racial minorities experience in their social lives is mirrored in the physical conditions of their communities as compared to many white communities. In communities of color across the country, from West Harlem, New York, to South Central Los Angeles, environmental degradation is but a symbol of the economic and social distress these communities suffer.

The physical decay of minority neighborhoods leads to the further balkanization of communities along racial lines. Much of today's white flight is driven largely by quality-of-life concerns such as environmental conditions, not the racism of three and four decades ago, though such racism persists. White urban refugees typically claim they are looking for a sense of community, place, and personal safety, each of which is in large part a function of the physical environment. The exodus of whites to new suburbs and remote rural areas has a double adverse social effect. It contributes to the increasing homogenization (in 1997, the forty fastest growing rural counties were between 70 and 85 percent white) and environmental degradation of rural America, especially in the Rocky Mountain states, upper Midwest, and New England.[69] And it drains urban centers of the cultural diversity and stable tax base that sustains them, perpetuating environmental and social decay.

Environmental injustice stemming from the disproportionate impacts of environmental harms on communities of color is integral to a degenerative social and environmental cycle. Racial inequality invites environmental harm, which perpetuates racial strife and polarization, which in turn exacerbates racial inequality.

Socioeconomic Equality

In terms of environmental degradation, the economic gap between rich and poor in the United States manifests itself in the physical separation between the haves and the have-nots. The same economic forces that drive the exploitation of resources, from the clear-cutting of forests for wood products and paper to the development of wetlands to build shopping malls, also enable the wealthy and professional class to distance themselves geographically from the adverse impacts of that exploitation. Mimicking the parallel nineteenth-century traditions of homesteading on the one hand and estate building on the other, today's affluent Americans are highly mobile, with the ability to move from community to community, and eager to establish exclusive, often gated, communities, sealed off from the noise, congestion, and blight that afflict many lower-income and minority neighborhoods.

The political and economic power of wealthy Americans enables them to develop and attract the resources and services necessary to protect their environment and to mitigate those environmental impacts that inevitably affect them. Environmental hazards such as ozone smog, polluted runoff from roadways, and lead contamination confront all communities, no matter how financially well off; it is simply a matter of degree and defense. Wealthy Americans have access to the financial, medical, political, and legal services that can help diminish or defend against those harms. They can move away from them, obtain high-quality medical care to prevent or treat the illnesses that result from them, lobby for more aggressive enforcement of environmental laws to abate them, or hire lawyers to sue those who cause them.

Meanwhile, lower-income communities, especially those of people of color, without the variety of resources available to their more affluent neighbors get saddled with a disproportionate share of environmental burdens while lacking both the tax base to pay for public goods like aggressive environmental enforcement, comprehensive health care, and physical infrastructure improvements such as mass transit, and the personal wealth to pick up and leave. Moreover, urban disinvestment has left many inner-city residents without adequate employment opportunities or access to mass transit to commute to jobs in outlying suburbs.

"One of the primary means by which individuals improve their life chances—and those of their children—is by moving to neighborhoods with higher home values, safer streets, higher-quality schools, and better services," Douglas Massey and Nancy Denton write in *American Apartheid: Segregation and the Making of the Underclass*, "Barriers to spatial mobility are barriers to social mobility, and by confining [lower-income people of color] to a small set of relatively disadvantaged neighborhoods, segregation constitutes a very powerful impediment to . . . socioeconomic progress."[70]

Over a century ago, the urban planner Frederick Law Olmsted attempted to lessen the physical segregation and social antagonisms between rich and poor in America's industrial cities by designing public parks that enabled people from all social classes to gather and commune with nature, especially those who could not afford to travel to the countryside. Olmsted believed that the experience of nature found in public parks had a democratizing effect on people, easing their private tensions while nurturing their sense of the common good. Olmsted's vision of an egalitarian society nourished by public parks captures the essential connection between socioeconomic equality and equal access to a high-quality environment. Unfortunately, Olmsted's vision, as relevant today as it was more than a century ago, has been undermined by widespread environmental degradation and persistent class division.

Public Investment and Privatization

Accompanying the growing gap between rich and poor in the United States has been the insidious trend toward public disinvestment and privatization. As more and more Americans insulate themselves within exclusive residential enclaves, seemingly far removed from the vicissitudes of ordinary life, they are also fortifying themselves with the services and facilities that enable them to preserve their autonomy and sustain their high standard of living. Thus, increasingly over the 1990s, wealthy Americans have pulled their children out of public schools in favor of expensive private schools, hired private police forces, erected gates around their property, and otherwise shut themselves off from the rest of the community. According to urban planners Edward Blakely and Mary Gail Snyder, these Americans are in pursuit of maximum control, protection, and privacy in their lives, shielded from the vagaries of chance and daily life.[71]

This trend has percolated up to the halls of the U.S. Congress and many state legislatures, where over the last twenty years, under the sway of right-wing ideologues, government investment in such public goods as mass transit, public education, welfare, and environmental protection has eroded. In the name of local governance and the free market, the White House, Congress, governors, and state legislatures have greatly reduced the amount of funding relative to expenses available for public goods, relying on municipalities and the market to take up the slack. Yet in many communities, and especially lower-income and minority neighborhoods, local government and the market have failed to follow through, lacking the revenues and the will, respectively, to meet their needs.

The trend toward privatization is evident, for instance, in the recent move away from publicly funded welfare programs to transitional assistance strategies, where welfare recipients are taken off the dole within a fixed amount of time and left to seek employment with minimal state involvement. Environmental protection is another area of social policy where there has been a shift from public to private solutions. As we will discuss in chapter 3, environmental protection programs are increasingly being retooled to incorporate market-oriented and private sector strategies in the place of the traditional command-and-control approach. Although these social policy developments appear to offer benefits in terms of effecting positive outcomes, they come at the expense of public commitments and democratic accountability, embodied in our public institutions. By its very nature, privatization signals a flight from government as the primary instrument of the common good.

As so-called big government has gone the way of the dinosaur, other institutions have stepped in to fill the void. In the 1980s and 1990s, the surrogate for big government has been big business. As a result of the frenzy of mergers and acquisitions over the past two decades, megacorporations—from banks and publishing houses to Hollywood movie studios and Silicon Valley software firms—have assumed the mantle of bigness once reserved for government. Yet as Benjamin Barber warns, such privatization "pretends to save government from its top-heaviness by sliding down the scale and empowering the local and parochial. In truth it only shifts power from the public to the private sector, leaving it as centralized and hegemonic as before, but liberated from democratic constraints like elections. . . . Privatization is not about limiting government; it is about terminating democracy."[72]

Disinvestment and privatization thus lead to a deficit in public commitment and accountability. They have an enervating effect on the ability of communities to evaluate and address public concerns, as individuals and institutions alike are preoccupied with such private matters as the ups and downs of the stock market, corporate mergers, and the like. Like any other resource, human attention is a limited commodity; the more of it that is devoted to private concerns, the less of it that is available for public concerns, including environmental protection.

The rise of massive corporations and the spread of market capitalism in the United States and abroad have driven unprecedented economic growth in the past half-century, and especially the 1990s. Such growth has proved indifferent to environmental impacts. Just as the emergent capitalist economy of the eighteenth and early nineteenth centuries decimated much of the natural resources and indigenous cultures of New England, denuding over 80 percent of its forests, wiping out dozens of wildlife species, and dislocating or destroying tens of thousands of native peoples, so too has the modern economy resulted in pervasive negative environmental effects, from countless thousands of toxic waste sites and polluted rivers and streams, to air pollution and habitat destruction wrought by unchecked sprawl and our ever-growing dependence on automobiles. These are among the more obvious environmental consequences of the unbridled economic growth born of a weak civic democracy, where the absence of an active, engaged citizenry creates a social and political void readily filled by shortsighted decisions or, worse, outright environmental assaults. As Daniel Kemmis remarks, the problem of corporate power and rapacious economic growth boils down to the "public's problem, stemming from its own lack of a clear identity. That lack of identity, in turn, stems from our overall failure to demand of ourselves an active practice of citizenship."[73]

Recently several progressive economists and social scientists have called into question conventional notions about the concept of economic growth. In a 1995 article in the *Atlantic Monthly*, Clifford Cobb, Ted Halstead, and Jonathan Rowe point out that a significant gap exists between what economists choose to measure in determining the gross domestic product and what Americans actually experience. They claim that traditional economic indicators "tell us next to nothing about what is actually going on."[74] The GDP, they explain, simply measures market activity in terms of the amount of money that changes hands; it does not differentiate between desirable and undesirable transactions. That is, it portrays transactions that result in the breakdown of the social structure and natural habitat as economic gain.

So, for instance, according to the GDP, "the happiest event is an earthquake or hurricane. The most desirable habitat is a multi-billion dollar Superfund site. All these add to the GDP, because they cause money to change hands."[75] Pollution represents more than just a one-time economic gain; it counts at least three times: once when it is produced (out of a smokestack at an automobile manufacturing plant, say), again when the nation spends money to clean up air pollution, and a third time when individuals spend money for medical bills for illness associated with dirty air. Thus, the GDP, our standard measure of economic growth, actually values the depletion of our natural resources. As the economist Herman Daly has despairingly remarked, the GDP treats the earth as if it were a business in liquidation.

Economic growth is a matter of civics as much as it is a matter of economics. To the extent that community members—from corporate managers and professionals, to workers, environmentalists, and elected officials—do not work together in identifying and preventing the adverse environmental impacts of economic growth, or are not disposed to care about the environmental consequences of their actions, such impacts will continue to undermine the civic and environmental health of communities. Environmental degradation resulting from development occurs *because of the lack of consciousness, participation, and planning* on the part of the diverse stakeholders in that development. Whether it is a matter of persuading corporate boards of directors to change their decisions, lobbying elected officials to pass new regulations protecting public health and the environment, or establishing citizen environmental committees to monitor and enforce against polluters, the level of civic engagement in communities can directly determine environmental outcomes.

From Civic Consciousness to Environmental Health

As America's civic health declines, the negative environmental effects of development, production, and consumption continue to mount, bolstered by the aggressiveness and power of corporations and the endurance of liberalism's embrace of private property and individualism. If our review of civic and environmental conditions teaches us anything, it is that human decisions determine environmental outcomes, for good or ill, and those outcomes shape the quality of our lives. This lesson defines the connection between civic life and environmental change and compels us to ask the questions, How do we arrive at decisions that best accord with our notions of civic and environmental quality? How do community members develop a civic consciousness and social infrastructure that will ensure effective and equal environmental protection for all and reverse the degenerative cycle of environmental and civic degradation that still controls?

In the next chapter, we will try to answer these questions by looking at the American environmental movement and its efforts to promote healthy and sustainable communities in the United States. Environmentalism is a uniquely powerful and appropriate tool for creating the kind of communities most Americans strive for but few enjoy. The environment itself is the ultimate common ground, the meeting point where people from diverse backgrounds and perspectives can come together to realize their shared interests in survival, health, and quality of life. It is the consummate rallying point, the reason and occasion for social change and changes of consciousness.

3

The Land That Could Be: American Environmentalism and the Pursuit of Sustainable Communities

The future of American environmentalism . . . remains outside the Beltway in the hands of a new civil authority forming in and around thousands of watersheds, forests, factories, and communities scattered throughout the country.
—Mark Dowie, *Losing Ground*

Poverty is hierarchic, smog is democratic.
—Ulrich Beck, *Risk Society*

Environmentalism as American Democracy's Emblem and Practice

Environmentalism is a uniquely potent democratic idea. It celebrates the ideal of community and interconnectedness that is at the heart of democratic theory. In the words of John Dewey, democracy is "the idea of community life itself." Environmentalism embraces Aldo Leopold's compelling land ethic—the notion "that the individual is member of a community of interdependent parts. . . . A thing is right when it tends to preserve the integrity and stability and beauty of the biotic community. It is wrong when it tends otherwise."[1]

The very concept of ecology is grounded in the notion of community. The term derives from the Greek words *oikos,* or "house," and *logos,* which means "account" or "study." Ecology thus denotes the practice of studying the house—literally, the place where we live, our community. Environmentalism also embraces community action, reflected in the famous injunction, "Think globally, act locally."

The environment itself is a traditional symbol of American democracy. National icons such as the bald eagle and institutions like the national park system, the world's first, are evidence of the deep connection between

American democracy and the country's natural bounty; to many, they are America's "best idea."[2] Recall that Jefferson and Tocqueville believed that the exceptionalism and success of American democracy rested on the availability of vast amounts of fertile land, which would breed industriousness, civic virtue, and a sense of freedom in the American people. For the poet Langston Hughes, America's physical landscape singularly embodied the ideals of equality and justice that, regrettably, the society itself failed to honor. In "Let America Be America Again" he exhorted,

O, let America be America again
The land that never has been yet
And yet must be—the land where every man is free
. . .
We, the people, must redeem
The land, the mines, the plants, the rivers
The mountains and the endless plain
All, all the stretch of these great green states
And make America again![3]

Further, environmental issues elicit democratic sentiments. Whether national parks or ozone smog, clean water or toxic waste, environmental issues are ultimately public issues that defy the boundaries of geography, politics, race, class, and gender. The organic act establishing the national park system, for example, provides that the parks were created for all people to enjoy, including future generations, no matter where they live, their party affiliation, skin color, income, or sex. Similarly, environmental problems like air pollution and hazardous waste inevitably affect entire communities, not just a handful of individuals. As political scientists Marc Landy, Marc Roberts, and Stephen Thomas explain, environmentalism "raises fundamental issues about who we are and what we care about."[4]

Yet despite its expansive democratic dimensions and potentially pivotal role in promoting civic health, American environmentalism has not prevailed as a democratic movement or social vision. This is because environmentalism and the environmental movement over time have mirrored the same kinds of civic and democratic problems that have affected the society as a whole, rendering environmentalists unprepared and unable to address the civic and social issues that are key to effecting long-term environmental results. As a result, the movement's successes, though considerable, have been limited or, worse, fleeting. Environmental degradation, evidenced in the harms that continue to affect natural resources and human health and that blight entire communities, is still a major threat to the long-term sustainability of American society. As we have seen, civic health is a necessary precondition to environmental quality. American environmentalism must therefore ultimately be measured against the yardstick of civic democracy.

The history of the American environmental movement reveals its partiality and shortcomings, first in what I call its romantic-progressive phase, and later in its mainstream-professional phase. Though chronologically distinct, these two phases are overlapping. Historians and observers of American environmentalism have tended to parse the movement into two separate epochs: the conservation-preservation era of the late nineteenth and early twentieth centuries and modern environmentalism originating in the 1960s and still with us today. Although such an approach provides a helpful analytical tool, we must nevertheless view the movement as a continuum to capture the ideas and attitudes deeply embedded in its institutions, membership, and agenda. The early history of the movement established the ideological and sociological framework on which modern mainstream-professional environmentalism is based. They are of a piece, united by the dual disposition of romanticism and professionalism that has shaped the entire history of American environmentalism.

Romantic-Progressive Environmentalism

As early as the mid-nineteenth century, George Perkins Marsh and Henry David Thoreau, among others, called for the conservation of nature, despairing, as Marsh did, that "Man has too long forgotten that the earth was given to him for usufruct alone, not for consumption, still less for profligate waste."[5] Thoreau lamented the rapaciousness with which Americans had exploited the land. "For one that comes with a pencil to sketch or sing," he decried, "a thousand come with an ax or rifle."[6] Reaction to the widespread settlement and exploitation of the land in the mid- to late 1800s, exemplified by massive timber harvesting, overgrazing of livestock, land speculation, and boom-and-bust mining, gave rise to the development of two distinct efforts aimed at protecting natural resources: preservation and conservation.

Spearheading the push for preservation was John Muir, the Scottish-born mountaineer who in 1892 founded the Sierra Club. The most vocal advocates for the creation of national parks like Yellowstone in 1872, the nation's first, and Yosemite in 1890, Muir and his fellow preservationists sought to protect wild nature from the harmful effects of human settlement and consumption. They opposed the notion that nature existed for the benefit of human use and instead upheld the environment's aesthetic and spiritual values as paramount. "Everybody needs beauty as well as bread," Muir declared, "places to play in and pray in where nature may heal and cheer and give strength to body and soul alike."[7] For preservationists, wilderness was the antidote to the materialism and arrogance of industrial society. Muir wrote, "Much is said . . . about 'the greatest good for the greatest number,' but the greatest number is too often found to be number one."[8] In promoting the protection of natural bounty, preservationists looked to the federal government as the most effective steward of America's parks and wilderness areas. In addition to their romanticism, preservationists were hard-nosed realists in their support of aggressive government oversight of public lands by the Department of the Interior and other natural resource agencies.

Preservation celebrated the sublime and the primitive in nature. Describing Yosemite Valley in California, Muir rhapsodized that it was "so beautiful that one is beguiled at every step, and the great golden days and weeks and months go by uncounted . . . [with] five hundred miles of waterfalls chanting together. What a psalm was that!"[9] Wild nature was a sacred temple where, William Cronon explains, "one had more of a chance than elsewhere to glimpse the face of god." Whereas in the eighteenth century wilderness was seen as the devil's playground, a frightful and forsaken place, by the end of the 1800s, Cronon explains, wilderness "became a place not just of religious redemption but of national renewal, the quintessential location for experiencing what it meant to be an American."[10] For preservationists, wilderness, and especially the vanishing frontier, was the authentic American landscape, the source of America's identity and salvation.

Preservationists embraced wild nature as the perfect counterpoint to the perceived ugliness and artificiality of nineteenth-century urban, industrial society. For them, wilderness embodied the romantic ideals of sublime grandeur and primitive simplicity celebrated by such European literary and intellectual lights as William Wordsworth and Jean-Jacques Rousseau. Although preservationists like Frederick Law Olmsted, Jr., Horace Albright, Stephen Mather, J. Horace McFarland, and Robert Marshall were overwhelmingly urban bourgeois, they considered modern urban industrial society degenerate. Parks and other undeveloped areas thus became a kind of tonic for preservationists—a refuge from the filth and bustle of urban life to be enjoyed through recreational activities like hiking, climbing, birdwatching, hunting, and fishing. Its cast decidedly antiurban and middle-class, preservation was a thoroughly romantic movement, an aesthetic reaction to the dramatic social and economic changes of the nineteenth century brutally manifested in the American landscape.

Simultaneous with the emergence of preservation, conservation arose in the 1890s, inspired by the progressive ideals, born of the Enlightenment, of rationality and science. Led by Gifford Pinchot, the country's first professionally trained forester (who helped found the Yale School of Forestry) and chief of the Forest Service under President Theodore Roosevelt, conservation responded to the environmental problems brought about by economic growth, specifically, the destruction of forests encouraged by cheap land prices and the overdevelopment of fragile water supplies, especially in the West. Pinchot believed that "the first duty of the human race is to control the earth it lives upon."[11] As chief forester, he opposed the preservation of forest lands, explaining that "the object of our forest policy is not to preserve forests because they are beautiful . . . or because they are refuges for wild creatures . . . but . . . the making of prosperous homes."[12] Much to the chagrin of preservationists like Muir, Pinchot went so far as to try to bring the national parks, a branch of the Interior Department, under the jurisdiction of the Department of Agriculture's Forest Service in hopes that park resources might also be developed, but to no avail.

Contrary to the goals of preservation, conservation aimed to use natural resources in the service of sustainable economic growth. Pinchot propounded an environmental vision that came to be known as progressive conservation. Joining the concept of economic efficiency with the goal of long-term yield of natural resources, Pinchot fashioned an environmental movement grounded in progressive ideology. He declared:

> The first great fact about conservation is that it stands for development. There has been a fundamental misconception that conservation means nothing but the

husbanding of resources for future generations. There could be no more serious mistake . . . [for] conservation demands the welfare of this generation first, and afterwards, the welfare of the generations to follow. . . . The outgrowth of conservation, the inevitable result, is national efficiency.[13]

Conservation held that environmental problems should be reduced to business issues and resolved using corporate tactics, such as centralized administration and scientific management institutionalized in expert agencies. During the presidency of Theodore Roosevelt, an avid outdoorsman and naturalist who, as a student at Harvard had written passionately about the closing of the American frontier, Pinchot's conservation reached its pinnacle. Pinchot profoundly influenced Roosevelt. For example, in 1908 Roosevelt convened the first Governors' Conference on Conservation at the White House and later established the National Conservation Commission. Roosevelt shared Pinchot's belief that expert government agencies should oversee the management of natural resources. Along with Secretary of the Interior James R. Garfield, Roosevelt and Pinchot orchestrated the nation's first conservation policy based on the principles of expertise and efficiency.

The tension between Muir's preservation and Pinchot's conservation came to a head beginning in 1908 with the fight over development of the Hetch Hetchy Valley in Yosemite National Park. Pinchot wholeheartedly supported San Francisco mayor James Phelan's plans to build a dam and reservoir in the valley to provide a reliable water supply to the booming Bay Area metropolis. For Pinchot, the Hetch Hetchy reservoir was an appropriate public use of federal lands. Muir and his preservationists denounced the proposed project as grossly utilitarian and materialistic, undermining the beauty and aesthetic value of the valley and Yosemite as a whole. Eventually, in 1913, the dam project was approved by the Congress and signed into law by President Wilson. Although conservation had triumphed over preservation, the Hetch Hetchy controversy established an enduring paradigm of environmentalism as comprising both romantic and progressive elements that would shape the future of the movement for decades to come.

Preservation and conservation are the bifurcated root system of modern environmentalism. Preservation endowed environmentalism with a romanticism that is one of its most hallowed hallmarks. Perhaps there is no greater figure in the history of American environmentalism than John Muir, a testament to the force of preservation's aesthetic and spiritual message. Preservation gave to environmentalism its bent to the sublime and transcendent embodied in wilderness areas and national parks while turning the face of environmentalism away from cities and other densely populated places (much like Jeffersonian agrarianism). As well, preservation's constituency, from Muir and the American Civic Association president J. Horace McFarland, to the entrepreneur Stephen Mather and landscape architect Frederick Law Olmsted, Jr., branded environmentalism with the stamp of the educated upper middle class.

Conservation introduced professionalism and expertise to environmentalism. Steeped in progressive ideology, conservation embedded in environmentalism a commitment to science, efficiency, and public administration. It gave rise to a new cadre of environmental professionals, armed with specialized degrees in resource management and public policy, and, with the preservation movement, promoted government as the appropriate steward of America's natural resources. As a decision-making model, conservation was decidedly top-down and professional.

Moreover, as author Robert Gottlieb explains, conservation's emphasis on expertise and rational management was eventually embraced "by the resource-based industries and other industrial interests attracted to the concepts of efficiency, management, and the application of science to industrial organization."[14] In other words, conservation opened the door to industry, the most powerful agent of environmental harm, establishing a comfortable alliance between environmental protection advocates and capitalists that would come partly to define the movement in later decades. Personified by such figures as Pinchot and Roosevelt, conservation also became associated with wealth and privilege, in line with preservation's bourgeois heritage, a socioeconomic legacy the two bequeathed to modern environmentalism.

The romantic-progressive thrust of preservation and conservation determined the basic contours of modern environmentalism and delimited its scope in terms of certain key social issues. Despite the efforts of Robert Marshall, a preservationist who cofounded the Wilderness Society and promoted the idea that social equality was central to wilderness protection, justice and democracy were left out of the romantic-progressive agenda. Many Sierra Club chapters, for example, deliberately excluded minorities from membership until the 1960s. In addition, the parks and nature preserves that environmentalists sought to protect were often off-limits to minorities and immigrants.[15]

Further, the romantic-progressive ideology shunned both urban areas and lower-income communities as appropriate subjects of environmental protection efforts. In fact, as Gottlieb suggests, the "anti-urban attitudes of the preservationists were . . . linked to their attitudes about class."[16] Cities were viewed by preservationists like Muir as places of squalor, pollution, and degeneration brought about by industrialization. They were also home to immigrants and minorities, who made up the workforce that fueled industrialization and were excluded from the ranks of the preservation-conservation establishment. Although conservation was primarily concerned with economic growth, this concern was directed not toward working-class Americans but to managers and professionals, the captains of America's thriving, resource-intensive industries. The focus on expertise, on the one hand, and aesthetic recreation, on the other, ensured that the romantic-progressive model would ignore minorities and lower-income Americans. Blended into the romantic-progressive model, therefore, was a nativistic and elitist disposition toward working-class Americans and minorities and the places where they lived.

Additionally, in its reliance on government and experts to solve environmental problems, the romantic-progressive model largely excluded laypeople and civic institutions from the environmental establishment. Despite the involvement of groups like the American Civic Association in the preservation movement, preservationists and conservationists alike failed to organize sizable citizen constituencies. Although organizations like the Sierra Club and National Audubon Society relied on local chapters for membership and activism, grass-roots citizen action would eventually be overshadowed by the centralized structure and professionalism of modern environmental organizations.

Grass-roots environmentalism developed on a separate track in the nineteenth century, in cities and rural areas, grounded in countless local struggles against industrial polluters and unwanted development. Although the urban environment and grass-roots activism were marginalized by the preservation and conservation movements, environmental issues related to cities and social justice did not go unaddressed. In the late 1800s, the social reformer and pioneering urban environmentalist Alice Hamilton, for example, took on the problems of industrial disease and occupational hazards, including phossy jaw (a disease affecting mine workers) and lead poisoning, during the early decades of the twentieth century. Similarly, Jane Addams, founder of Chicago's Hull-House Settlement in 1888, promoted sanitation and public health on behalf of the city's immigrant and minority neighborhoods and helped organize local citizens in grass-roots reform efforts. Hull-House's mission was to promote social democracy and community revitalization through a mixture of neighborhood organizing, professional advocacy, and technical assistance. Still others, like the planner Benton MacKaye, sought to improve the living and working conditions of urban residents through regional planning aimed at better integrating the natural environment into cities.

Emerging parallel to the preservation and conservation movements, grass-roots efforts oriented toward cities and citizen activism comprised a legitimate environmental agenda in the early twentieth century, enriched by an emphasis on social democracy. Yet they remained outside the focus of the romantic-progressive model and would remain marginal to the mainstream-professional environmentalism that arose later in the century.

Mainstream Professional Environmentalism

Firmly established by the 1920s, the preservation-conservation model of environmentalism determined the course of environmental policy and activism for several decades to follow. With the expansion of the federal government under President Franklin Roosevelt in the 1930s, the Departments of Interior and Agriculture developed a host of policies related to resource management and protection of parks and wilderness areas focused on providing for the growing material needs of the country while protecting pristine nature reserves for recreation. Interrupted by World War II, federal environmental policy continued to evolve in the 1950s, still aimed at resource use and wilderness protection, but with a particular emphasis on the emerging issues of population growth, economic expansion, and technological innovation, including nuclear energy.[17]

With the development of major public works projects in the West, widespread suburbanization, and a handful of well-publicized controversies concerning industrial pollution in the 1950s and 1960s, the modern environmental movement began to crystallize. Government-sponsored energy and water projects during this period, including a proposed nuclear power plant at Diablo Canyon in California, and proposed dams and hydroelectric facilities at Echo Park in Utah's Dinosaur National Monument, and the Glen and Grand Canyons in Arizona, riled the preservationist forces of groups like the Sierra Club and the Wilderness Society, which waged heroic battles of unprecedented scale in opposition to what they saw as catastrophic environmental assaults. In 1964, the Wilderness Act was passed, a milestone in the struggle for legal protection of the nation's backcountry.

During this period, notes historian Kenneth Jackson, Americans began migrating to the suburbs as never before, leaving behind the deteriorating cities with help from new federal mortgage subsidies, highway infrastructure, and housing subdivisions.[18] As the suburbs grew, smog, traffic, and sprawl soon followed, provoking a new environmental consciousness among suburbanites. Concurrently, public understanding of the physical and biological issues underlying environmental harms grew as scientific information became more widely disseminated, and media outlets began covering environmental stories. This, too, helped generate a new awareness of the natural world among a well-educated and affluent suburban constituency.

Its political consciousness forged by the New Deal and World War II, this new constituency looked to government for answers to public problems and therefore saw public policy as a legitimate vehicle with which to address environmental issues. Meanwhile, with the publication of Rachel Carson's *Silent Spring* in 1964, warning of the severe hazards to wildlife and human health resulting from the use of industrial chemicals like DDT, Americans in general, and especially suburban environmentalists, took up arms in defense of endangered species and wilderness, environmentalism's age-old foils to industrialization.

Picking up on the romantic-progressive tradition born nearly a century earlier, the modern environmental movement was quickly transformed into what some observers have called "the secular religion of the white middle-class."[19] With the first Earth Day in 1970 and graphic media coverage of environmental horrors like oil spills and belching smokestacks, the educated upper-middle-class environmental constituency persuaded the Nixon administration in the early 1970s to erect the building blocks of the nation's modern environmental law and policy system, thus marking the arrival of mainstream-professional environmentalism.

Oriented toward professionalism, law, and science, environmental law and policy became the stomping grounds of a legal-technical elite drawn not only from newly formed government agencies such as the EPA and Council for Environmental Quality, both created in 1970, but from established environmental groups like the Sierra Club, Wilderness Society, and National Audubon Society, and start-up nonprofit environmental law organizations like the Environmental Defense Fund, Natural Resources Defense Council, and Sierra Club Legal Defense Fund. Flip sides of the same coin, government and public interest environmentalists, in the spirit of their preservationist and conservationist forebears, came to shape the movement in last three decades of the twentieth century.

The EPA and the Environmental Law and Policy System

Betraying their romantic-progressive heritage, modern environmentalists looked first to government for solutions to the nation's environmental problems. As a result of intensive lobbying on the part of environmental and consumer advocates such as Ralph Nader and the leadership of politicians like Maine senator Edmund Muskie, a new set of environmental laws and policies was enacted in the early 1970s, including the National Environmental Policy Act (NEPA) and Clean Air Act in 1970 and the Water Pollution Control Act (Clean Water Act) and Federal Fungicide, Insecticide, and Rodenticide Act in 1972.

With the passage of these and other environmental laws, EPA's work was cut out for it. As the government nucleus of mainstream-professional environmentalism, EPA helped define much of the character of modern environmentalism. It served instantly to legitimize environmental issues as a major public policy discipline by consolidating specialized resources and staff within a large bureaucratic institutional structure beyond anything that had existed before. Accordingly, environmentalism became mainstream in the sense that it assumed an accepted place in government and the society at large. Moreover, as a regulatory agency comprising mainly lawyers, engineers, and economists, EPA reflected the professional strain of environmentalism inherited from the romantic-progressive model.

Broadly defined, EPA's mission is to protect the public health by setting nationally uniform standards for air and water quality, emissions of harmful pollutants, and handling of hazardous substances. To obtain permits from state and federal regulatory agencies, polluters must demonstrate that they will comply with these standards. Generally environmental regulations do not allow the costs of compliance to be considered in permitting decisions. As a matter of law, therefore, public health is paramount. However, environmental regulations are not designed to *prevent* pollution per se, but only to *control* it according to health-based standards. Implicit in the environmental law and policy system is thus the notion that some level of pollution is acceptable; pollution control, not pollution prevention, is the summum bonum of environmental protection.

As the nation's leading environmental agency, EPA oversees all state environmental agencies, reviewing and approving state environmental programs and establishing a baseline of pollution standards that each state must match or exceed. Federal environmental law thus preempts state and local authority, setting standards and procedural requirements to guide state and local actions. Before the establishment of federal environmental laws, states and local governments regulated pollution as a nuisance under their police powers. In addition, common law courts entertained nuisance actions by plaintiffs who claimed injury from nearby polluting facilities, and the federal government assisted states with research, technical support, and funding for water treatment plants and other environmental projects.[20]

Like all other government agencies, EPA has a democratic responsibility to use its expertise "to frame questions so that public debate can be made coherent and intelligible . . . [and to] tease out the essential social and ethical issues from the welter of scientific data and legal formalisms in which those issues are enveloped."[21] However, this role has proved extremely difficult to perform. Although the public has always held the power to influence EPA decision making, the regulatory and scientific focus of EPA's mission and its dense professional culture have tended to inhibit public influence wielding and involvement. Employing several different kinds of professionals, each with their own specialized discourses, EPA, as political scientists Marc Landy, Marc Roberts, and Stephen Thomas suggest, has often bred "misunderstanding and conflict" among its stakeholders, including the public.[22] With lawyers, engineers, and economists as the dominant players within EPA, the agency has historically shied away from its role as public educator and promoter of democratic decision making and instead has typically presented itself as a legal-technical elite above the fray of politics and public pressure.

This image was established early in EPA's history, when the agency's Office of General Counsel held in 1970 that, due to the scientific bent of environmental law, civil rights laws and other policies addressing social and political issues were not relevant to its mission. EPA's first administrator, William Ruckelshaus, echoed this position when he argued before the U.S. Commission on Civil Rights in 1971 that EPA's scientific role in setting environmental standards precluded the application of civil rights law to environmental programs.[23] EPA has thus historically given priority to its responsibility to science and law over its obligation to ensuring equal environmental protection and democratic decision making.

The Environmental Law and Policy System The laws and programs that EPA is charged with implementing and enforcing have not made its democratic role any easier. Grounded in a complex regulatory structure designed to deal with environmental problems one at a time and pollutant by pollutant, the laws and regulations that comprise federal environmental law and serve as a template for all state environmental programs are, environmental law professor E. Donald Elliot explains, "premised on the fiction of an omniscient center" capable of dealing with all environmental problems in a centralized and uniform manner, commonly referred to as command-and-control regulation.[24] This pollutant-by-pollutant, medium-by-medium approach, dividing up problems so as to make them more soluble and accessible, has resulted in fragmented regulations that, most observers agree, have failed to address the degradation of ecosystems as a whole. Consequently, pollution problems are often merely shifted from one medium to another, or from one geographic location to another, with often marginal net environmental benefits.

In addition, environmental laws are designed to address only part of the life cycle of pollutants. Thus, for example, the Clean Air Act focuses on controlling pollution that comes out of a factory's smokestack but ignores the environmental impacts caused when materials used in a factory have already been produced at another location or when the finished product reaches a firm's customers. In other words, environmental laws are intended merely to control pollution, not prevent it. Emissions standards and permits are by definition allowances to pollute. Accordingly, the key regulatory issue is how much, not whether, pollution is safe and tolerable. Environmental law thus treats pollution as a necessary evil, to be controlled as opposed to eradicated.

A case in point is the pollution credits scheme under the 1990 Clean Air Act Amendments. Designed as an inexpensive, flexible, market-based alternative to command-and-control regulation, pollution or emissions credits allow companies to trade the right to pollute. Under the act, any company that reduces its emissions below federal standards can earn pollution credits that it can then sell to other companies struggling to comply with the standards. However, the act turns a blind eye to the regional environmental inequities that exist between upwind and downwind states. Environmentally sensitive areas in the Northeast like the Adirondacks and Berkshires have suffered significantly as a result of emissions credits, as polluters in the South and Midwest have found it cheaper to buy pollution credits (from the owners of newer, cleaner northeastern facilities, among others) than to reduce toxic emissions that eventually blow back into the Northeast and contribute to acid rain. Some observers have gone so far as to call emissions credits "immoral."[25]

Based on national standards, environmental laws like the Clean Air Act and Clean Water Act target the nation's largest pollution sources, such as power plants, refineries, and chemical plants, focusing on readily identifiable pollution points within a source (so-called point sources), such as smokestacks or outflow pipes. To a considerable extent, these laws have been successful in reducing pollution of the nation's air and water. Environmental policy observers like the journalist Gregg Easterbrook are right to point out, as he did in his 1995 book, *A Moment on Earth: The Coming Age of Environmental Optimism*, the significant strides the country has made in improving the overall state of the environment. A self-proclaimed environmental optimist, Easterbrook and other optimists celebrate the efficacy of the nation's environmental law and policy system while discounting the alarmism, pessimism, and stridency of environmentalists who claim the nation still has far to go before it can claim victory in the war against pollution.

Yet when the lens is shifted away from large polluters and environmental indicators like national air quality and focused more closely on smaller pollution sources and local environmental conditions, environmental optimism ineluctably dims. Smaller, nonpoint source polluters, like farms, dry cleaners, and auto body shops, as well as local land use decisions, often determinative of environmental outcomes like sprawl and loss of open space, are historically beyond EPA's and state environmental agencies' purview. EPA's jurisdiction is explicitly limited to permits and pollution control; it has no legal authority over local land use decision making, for example, rendering the agency largely powerless to deal with environmental harms resulting from shortsighted development decisions. As Donald Elliot points out, command-and-control regulation, while effective at dealing with large pollution targets like power plants, does not work well with more diverse and diffuse environmental problems and circumstances, such as ambient air pollution generated by automobiles, habitat destruction from overdevelopment, or asthma caused by the leaky roofs of dilapidated housing stock.

Even the potent citizen suit provisions found in most federal environmental statutes, giving local citizens the right to bring legal challenges against alleged polluters, are often useless given the limited scope of environmental laws and the highly technical and scientific nature of environmental litigation. As a result, local communities are often forced to resort to laws considered outside the regime of environmental law, such as land use ordinances, sanitation codes, and public health regulations, for relief from many types of environmental harms. Environmental laws like the Clean Air Act are simply irrelevant in many cases. The emphasis on large point sources in federal environmental laws has narrowed EPA's and states' jurisdiction over environmental harms, ignoring the myriad nonpoint source polluters and local environmental decision making that often constitute the most serious threat to environmental health.

Moreover, the uniform national standards approach, or command-and-control system, established by EPA and implemented by the states has prevented local communities from adopting environmental standards more responsive to local conditions. Commenting on the National Ambient Air Quality Standards (NAAQS) under the Clean Air Act, Leon Billings, former staff director of the Senate Public Works Commission, claimed that NAAQS were "the biggest mistake we ever made" because they took the issue of air quality out of the hands of local communities and put it on the national level, where technocrats and industry have the advantage and typically carry the day.[26] As a result, NAAQS do not adequately protect certain local communities, such as inner-city neighborhoods, where pollution hot spots occur due to the presence of multiple pollution sources and high exposure risks.

The Comprehensive Environmental Response, Compensation and Liability Act, known as Superfund, is another example of the limits of environmental law. Perhaps the country's most complex environmental statute, Superfund, enacted into law in 1980 and amended in 1986, was designed to identify the nation's most polluted sites and to force parties responsible for that pollution to clean it up. Under a joint and several liability scheme, in which a single party can be held liable for the total cleanup costs at a site even though that party minimally contributed to the site's contamination and release of the pollution was legal at the time, EPA pursues responsible parties in hopes they will voluntarily take steps to remediate their sites. Alternatively, EPA can go forward with cleanup on its own, using funds specifically designated for such purposes, and then sue the responsible parties for up to three times the cost of cleanup.

The heart of the statute concerns the science and process of hazardous waste site cleanup, setting forth the requirements for remedial investigations, feasibility studies, technology assessments, and the like. The key actors under Superfund are EPA lawyers and engineers and the responsible parties themselves; the statute offers little room for local community involvement other than the scant requirements of public notice of cleanups, comment on proposed actions, and public meetings in the areas near the sites.

The epitome of command-and-control regulation, Superfund is widely regarded as a failure; relatively few sites have been cleaned up under the program since it was established. Of the 1,359 Superfund sites (also known as brownfields) around the country, most of them in cities and old industrial areas, only 37 percent have been remediated after eighteen years.[27] Most of the action has been in the courts, where potentially responsible parties have sued both EPA and each other over the issues of liability and cleanup costs. More important, Superfund's liability scheme has erected a figurative iron fence around most contaminated sites, scaring off would-be developers and, as a consequence, directing them to clean, undeveloped sites, often in suburban and rural areas.

Superfund is an example of the shortcomings of not only the command-and-control approach but EPA's reliance on complex legal rules and science to solve what is for the most part a historical, social, and economic problem: the past contamination (often legal) of former industrial sites long since abandoned by businesses that have either been dissolved or moved to more economically viable areas.

Some Reinvented Approaches to Environmental Regulation Concern about the efficacy of command-and-control regulation, uniform standards, and fragmentation has led recently to a transformation in the way EPA and many of its state counterparts conceive and enforce environmental laws. In the name of regulatory reform and devolution of federal power first championed by right-wing and centrist politicians during the Republican congressional sweep in 1994, EPA and many state environmental agencies have initiated programs aimed at simplifying and streamlining environmental regulation to achieve better and faster environmental results than under the traditional command-and-control system. Emphasizing local control, non-adversarial problem solving, and economic incentives, EPA's new approach is an attempt to reinvent environmental law and policy.

For example, under Project XL (shorthand for "Excellence and Leadership"), EPA has invited firms to apply to participate in a select group of pilot facilities to engage in developing a "cleaner, smarter, cheaper" approach to regulating pollution at their plants. Project XL relies on site-specific agreements with polluters to develop alternative compliance techniques that can serve as national models. In exchange for regulatory relief in the form of waivers from permitting and reporting requirements for such things as new plants and production lines, firms agree to go beyond existing compliance strategies to achieve greater environmental performance.[28] For instance, a company might seek to remove wastewater sludge from a toxic waste register requiring special treatment and thus considerable expense. In turn, it might propose to direct those savings to on-site sludge drying and to recycling and reusing other waste that is currently sent to a landfill.[29]

In the view of Anne Kelly, a former special assistant to the regional administrator of EPA/New England who spearheaded a number of Project

XL pilot initiatives, Project XL aims to strike a balance between the need for efficient, flexible regulations and EPA's mandate to protect the environment and human health without compromise. She explains:

> Project XL is an experiment in seeing how far EPA, working with the regulated community and other key stakeholders like environmentalists and workers, can go in helping the private sector do the right thing. No one at EPA wants to give up the store, nor do we want to be perceived as pandering to corporations and industry. We simply want to try to advance the project of getting more and better environmental results by demonstrating that environmental performance is really a business advantage. But we first have to establish the models.[30]

EPA/New England, among a handful of other regional offices, has earned a solid reputation as being innovative and responsive in the face of the failures and unintended consequences of traditional environmental regulation, piloting a variety of new programs, including Project XL, meant to improve the state of the art. The regional administrator, John DeVillars, has helped push the agency in new and important directions, such as taking on the regional environmental and economic problems associated with sprawl and engaging the tricky land use and planning issues inherent in brownfields redevelopment. Notwithstanding EPA's limited jurisdiction, the New England office has successfully pioneered several creative strategies for dealing with new environmental challenges while respecting the limits of its statutorily defined role. Other EPA regional offices are following suit, attempting to adapt to the same kinds of changes that the New England region has faced.

At the state level, similar innovations are being tested. For example, the Massachusetts Department of Environmental Protection (DEP) has proposed its own version of reinvention, the Environmental Results Program (ERP). In an attempt to simplify the environmental permitting process for the roughly 10,000 companies considered to be lower-risk polluters (although together they generate annually 80 percent—170,000 tons—of the state's hazardous waste), thereby freeing up DEP resources to conduct inspections and enforce against the state's most serious environmental violators, the Massachusetts DEP is, in the words of former Commissioner David Struhs, "replac[ing] the command-and-control practice of 'engineering the permit' with a facility-wide, performance-based compliance self-certification program."[31] In essence, ERP lets the polluter design the compliance techniques and monitor compliance performance, with DEP oversight and guidance. Working on a pilot basis with dry cleaners and photo processors, two industries long neglected by environmental agencies, ERP also does away with the requirement of separate annual compliance reports for each category of pollution (air, water, solid waste) and instead requires a single multimedia summary.

As examples of the emerging models of flexible, results-oriented environmental regulation, Project XL and ERP represent innovative solutions to the problems of command-and-control regulation and fragmentation. As well, these programs deliberately extend beyond traditional environmental regulations in targeting firms such as dry cleaners that have until recently fallen outside the purview of federal and state environmental agencies. This is because, on an individual basis, such businesses are not considered significant polluters when measured against other industries. Yet, taken collectively, dry cleaners are a major source of toxic pollution, including perchlorethylene, a deadly gas.

Regulatory reinvention presents possible pitfalls when viewed from the standpoint of democratic accountability. For example, a Project XL agreement involving the Intel Corporation's Chandler, Arizona, plant has been roundly criticized by a coalition of environmental, community, and labor organizations as a sweetheart deal between Intel and EPA. The agreement provides that in exchange for implementing "Design for the Environment" production systems, among other environmental improvements, Intel will be allowed to increase its overall pollution emissions (in proportion to production output) and emissions of individual pollutants, with little public or government oversight.[32] According to environmental lawyer Sanford Lewis, the agreement was

> far weaker than agreements reached by community groups negotiating directly with corporations without EPA involvement. Elsewhere community groups had won commitments . . . to reduce pollution . . . and to target and eliminate use of toxics, even to achieve zero pollution discharge. And, looking at the process by which it was released, it became apparent that serious power imbalances allowed the company to obscure and override environmentally-related objections. In contrast, other communities' neighbors and workers had negotiated for and won a right to conduct detailed environmental and safety auditing of local plants, and then to negotiate at a balanced bargaining table on the range of actions to be taken by the firm to protect environment, health and safety.[33]

Compounding such concerns is a series of recent audits by EPA's inspector general documenting widespread failures by federal and local officials in several states to monitor and enforce basic requirements of the Clean Air and Clean Water Acts. The reports, issued in March 1998, point out that many polluting facilities operate under obsolete permits or none at all, inspectors fail to visit and review facilities' environmental compliance, and states are falling short of environmental goals. Fred Hanson, EPA's deputy administrator, stated that the audits revealed "a troubling trend of possible deficiencies, at least in some states."[34]

In Idaho, for example, five major facilities along the heavily polluted Boise River were found to be operating with permits that had been automatically extended years ago and were never updated. This situation appears to be the rule, not the exception, in the Northwest, where even when pollution discharges were known to violate environmental laws, EPA rarely took enforcement action. As a result of the audits, many observers are left wondering whether the states and EPA are dedicated to enforcing pollution laws and thus whether giving polluters more flexibility is simply passing the buck.

ERP too has been challenged by environmentalists who are concerned that streamlined permitting and self-certification of compliance performance would, according to Conservation Law Foundation attorney Rusty Russell, "stymie public oversight of toxic hot spots often found in low-income neighborhoods, as well as of thousands of small but collectively significant polluters scattered across" the state.[35] In addition, environmentalists question whether in fact DEP's resources will be redirected to target the biggest polluters. Many ERP critics believe that when it comes to pollution reduction, there is no substitute for the threat of enforcement of environmental laws. Even business executives acknowledge that this threat is one of the principal reasons companies undertake compliance measures.[36]

Success and Failure: The Lingering Gaps in Environmental Protection "Many of today's environmental problems are different from those tackled over the last several decades," environmental policy professors Daniel Esty and Marian Chertow write. "Harms such as ozone layer depletion, climate change, or endocrine disruptors are less plainly apparent, more subtle, and more difficult to address than the black skies or orange rivers of the 1960s."[37] The evolution of EPA and the environmental laws it implements reveals the constraints inherent in bureaucratic, top-down regulation and expert problem solving. Although EPA and the nation's environmental law and policy system in general have achieved a great deal in the way of environmental successes—consider the return of dozens of endangered species like the bald eagle from the brink of extinction and the dramatically cleaner air in the Los Angeles basin—pervasive pollution problems persist, owing in part to the system's centralized, fragmented, and rigidly professional approach.

Starkly absent from the environmental law and policy discourse over the past twenty-five years has been a focus on community-based or grass-roots strategies emphasizing pollution prevention and environmentally sound land use and planning decision making, issues that are integral to effecting long-term, beneficial environmental results. Moreover, in trying to move beyond the pollution-based and law-driven model of environmental protection by turning to industry for leadership and collaboration, EPA and state environmental agencies have risked further alienating local communities and a public long used to being left out of environmental decision making. Again, Sanford Lewis speaks to this issue:

> Instead of focusing on what kinds of "favors" can be granted to entice good corporate behavior, reinvention efforts can be better directed toward new rules and structures to encourage and foster environmental and community group initiatives, and to discourage corporations from using reinvention as cover for regulatory rollbacks. In short, the reinventors need to refocus the debate so that the relative roles of corporate and community power are addressed, front and center, in the reinvention debate. Until then, environmental law and policy may continue its slide down the slippery slope toward government by and for the corporation.[38]

The irony is that the push to reinvent environmental regulation is being driven in part by the recognition that local, multi-stakeholder involvement in environmental decision making is key to effecting better environmental results. In the light of the progressive origins of mainstream-professional environmentalism, it is not surprising that EPA and state agencies are often more successful in dealing with businesses than with local communities. As professionals and experts, government environmentalists are not used to reaching out to community stakeholders; indeed, they did not have to until recently.

Although there can be no doubt that the nation's environment is far better off on account of EPA's efforts over the past three decades, a yawning gulf exists between government-led environmental initiatives and community-based environmental planning and problem solving. There has been little room in the environmental law and policy system for community involvement and initiatives despite the local nature of most environmental problems. Consequently, while larger pollution sources, the traditional targets of environmental law, have for the most part been effectively controlled (save for those older facilities that are allowed by law to continue to pollute beyond regulatory limits and the occasional bad actor), myriad smaller sources, from auto body shops and housing subdivisions to highways and shopping malls, and the local decision making that spawns them, have continued to degrade local environments.

The Public Interest Environmental Establishment

Now let us turn to the other environmentalists: the advocates and activists from the public interest organizations who have grown up alongside their EPA and state agency counterparts. What have they accomplished in terms of promoting democratic environmental protection?

John Muir's Heirs Just as Gifford Pinchot and John Muir are the Janus-faced image of romantic-progressive environmentalism, so too are government agencies and public interest organizations the twin faces of mainstream-professional environmentalism. Like their government colleagues, public interest environmentalists appeared in force in the early 1970s, picking up on the tradition started by the Sierra Club in 1892 and gradually expanded by organizations such as the Wilderness Society, the National Wildlife Federation, and National Audubon Society in the mid-1900s. Led by professionals, especially lawyers, and deep-pocketed, well-connected boards of directors, groups like the Natural Resources Defense Council, the Environmental Defense Fund, and the Sierra Club Legal Defense Fund were founded on the belief that scientific and legal expertise, appropriately wielded, could solve environmental problems. Acolytes of Muir, this new breed of environmentalists was thoroughly romantic in their attitudes about the environment. As well, they were idealists about the power of law and science to effect positive social change and thus privileged professionalism as a tool for environmental advocacy. With offices in major metropolitan areas like Washington, D.C., New York City, and San Francisco, the new public interest environmental establishment focused on national issues while developing staffs of lawyers and scientists trained by the nation's elite universities.

Even groups like the Sierra Club and National Audubon Society, which historically relied on locally based chapters and societies and engaged citizens in direct action campaigns, moved to a more centralized, expert-oriented organizational structure in the 1970s and 1980s. The National Wildlife Federation, traditionally dominated by hunting and fishing groups active in state affiliates, also gradually shifted to a professional-based staff in the 1970s, expanding its mission beyond outdoor recreation and wildlife to include environmental protection generally.[39]

Mainstream-Professional Environmentalism Versus the Grass Roots From the start, the public interest side of the mainstream-professional movement directed its attention to federal environmental policy, shadowing the movements of EPA and the environmental law and policy system it spearheaded. The new environmental organizations were created explicitly to take advantage of the evolving regime of environmental laws enacted during the Nixon administration that gave citizens a solid foothold in the enforcement of environmental regulations. Through such techniques as public participation in the environmental impact review of construction projects under NEPA and legal challenges brought under the citizen suit provisions of a host of environmental laws, these groups were empowered by environmental statutes to act on behalf of their membership and other concerned citizens with the explicit purpose of protecting the environment.

Yet despite the seemingly democratic purpose of such legal tools as public participation and citizen suits, the public interest organizations have failed to promote widespread, bottom-up citizen involvement. Their strategy focused instead on the federal courts and Congress, pursuing litigation and legislative action that sought to broaden the scope of environmental regulations while aggressively attacking polluters with an assortment of legal weapons afforded by the Clean Air Act, NEPA, and other environmental statutes. Case after case was filed in the 1970s and 1980s, sometimes with stunning success, as in the famous case, *Tennessee Valley Authority v. Hill,* in which environmentalists halted operation of the Tellico dam on behalf of a then little-known fish called the snail darter. Through countless actions involving the Sierra Club, the Natural Resources Defense

Council, the Environmental Defense Fund, and other groups, public interest environmental litigation created a vast, complex body of common law rulings requiring a cadre of specially trained lawyers to interpret and deploy. As author Mark Dowie explains, mainstream environmentalism became a profession dedicated to constant "wrestling with government and corporations over laws and standards."[40]

Moreover, building their membership through direct mail solicitations instead of political organizing and direct action, the public interest organizations appealed to the white, middle-class, suburban constituency that blossomed during the 1960s but was not inclined to engage in hands-on activism. Whereas in earlier decades environmental organizations like the Sierra Club, National Audubon Society, and National Wildlife Federation were essentially decentralized, comprising active, albeit socially and racially homogeneous, local and state chapters, the public interest environmental organizations by the 1970s became centralized, in effect cutting themselves off from local constituencies. This centralization, made possible by direct mail support from a largely passive constituency, coupled with their focus on litigation and lobbying, ensured that the public interest environmental establishment would have few ties to local, grass-roots environmental efforts. In essence, the full-scale institutionalization of environmental protection in the form of a sophisticated administrative structure (EPA and state environmental agencies) and legal apparatus (the federal environmental law and policy system) dictated that the public interest environmental organizations act in kind, adopting the institution's professional approach and centralized structure.

Yet just as mainstream-professional environmentalists were cutting off their connections to grass-roots constituencies, community activists across the country began rallying around the cause of environmental protection in their neighborhoods, towns, and counties. In Love Canal, New York, Lois Gibbs organized her neighbors in the late 1970s to confront the industrial polluters who had turned her quiet, working-class upstate neighborhood into a toxic nightmare, poisoned by over two hundred chemicals. Gibbs started the Citizen's Clearinghouse for Hazardous Waste (now called the Center for Health, Environment, and Justice) in 1981 to assist other grass-roots groups fighting industry, government, and even mainstream environmentalists in the effort to clean up their communities. In Los Angeles, school teacher Penny Newman founded Concerned Citizens in Action in 1979 to demand the cleanup of the Stringfellow Acid Company pits in Glen Avon. Like Gibbs, Newman organized her neighbors in a comprehensive environmental campaign and got results. Through their efforts, Gibbs and Newman helped put the issue of toxic waste on the map and caused Americans to take a hard look at environmental conditions in working-class communities.

In eastern cities, urban residents organized throughout the 1970s to oppose major highway construction projects such as the inner beltway project in Boston in 1972. Fighting to save their neighborhoods in the face of demolition, Boston residents and other urban denizens established city-wide coalitions to promote urban environmental quality and their pride of place. The inner beltway, and several other transportation projects like it, never got off the ground thanks to the power of grass-roots environmentalists and their message of community preservation.

In 1982, black residents of rural Warren County, North Carolina, successfully organized public demonstrations in opposition to a proposed landfill for polychlorinated biphenyls (PCBs), leading to five hundred arrests and attracting national media attention. The event is credited with kicking off the national environmental justice movement, which throughout the 1980s and 1990s trumpeted the call for equal environmental protection for all Americans, regardless of income or color. Borrowing from the success of efforts like the Warren County protest, the Los Angeles–based Labor/Community Strategy Center initiated a massive campaign in the early 1990s involving thousands of "straphangers" to improve air quality and mass transit service in the city for lower-income residents, resulting in an overhaul of the city's transit system.

These and numerous other grass-roots environmental efforts over the past several decades represent the lasting legacy of the Hull-House movement. Yet despite their local successes, such campaigns have not significantly influenced national policy debates or the agenda of mainstream-professional environmental organizations. Often ad hoc and singular and always community driven, grass-roots issues and strategies have largely failed to percolate up to the higher echelons of the environmental establishment. The result has been a wide gulf between local environmentalists and their professional counterparts.

Many grass-roots groups have been openly critical of the mainstream-professional movement as unaccountable and elitist. They allege that the alignment of public interest environmentalists to the regulatory system and away from local constituencies caused them to lose their independence and contrarian identity in dealing not only with the regulators like EPA, but the regulated, that is, corporations.

For instance, the Group of Ten, the environmental leaders from the nation's ten largest, most powerful environmental organizations (the Environmental Defense Fund, the Environmental Policy Center, Friends of the Earth, National Audubon Society, National Wildlife Federation, the National Parks and Conservation Association, the Natural Resources Defense Council, the Izaak Walton League, the Sierra Club, and the Wilderness Society) who first came together in 1981 during President Ronald Reagan's first term to devise a unified strategy to counter Reagan's hostile stance toward environmental protection efforts, eventually joined with corporate leaders in the late 1980s in defense of market- and voluntary-based environmental policies like pollution credits. Already perceived by grass-roots environmentalists as elitist, the Group of Ten's embrace of corporate environmental leadership and controversial, market-oriented environmental policies had the effect of further undermining the credibility of the mainstream-professional organizations in the eyes of grass-roots groups, who for decades had been engaged in local environmental struggles against those same corporations. The public interest environmental establishment, embodied by the Group of Ten, thus came to be seen by many as undemocratic, an exclusive club of white men who, in the words of Mark Dowie, achieved environmental protection "through negotiation among the powerful," with little in the way of public involvement and accountability.[41]

The Environmental Justice Critique Perhaps even more than their perceived alliance with industry and government policymakers, the public interest environmental establishment's lack of racial and socioeconomic diversity, in terms of both its personnel and advocacy agenda, irked environmentalists from groups outside the mainstream. In the late 1980s and early 1990s, environmental justice activists publicly took issue with the mainstream-professional movement's overwhelmingly white, middle-class membership. Comprising activists from communities of color across the country, the environmental justice movement carried on the social justice tradition started by nineteenth-century activists like Jane Addams and, more directly, the civil rights movement of the 1950s and 1960s. Calling for equal environmental protection, environmental justice activists waged battles as early as 1982 against polluting facilities in places like Warren County, North Carolina; East Los Angeles, California; West Harlem, New York; Louisiana's Gulf Coast; Houston, Texas; and tribal reservations in the Southwest.

Environmental justice emerged as a powerful critique of the mainstream-progressive movement, alleging that American environmentalism had institutionalized environmental injustice for communities that were home to lower-income residents and people of color. As sociologist and environmental justice activist Robert Bullard explains,

> The current environmental protection paradigm has institutionalized unequal protection; traded human health for profit; placed the burden of proof on the "victims" rather than on the polluting industry; legitimated human exposure to harmful substances; promoted "risky" technologies . . . ; exploited the vulnerability of economically and politically disenfranchised communities; subsidized ecological destruction; created an industry around risk assessment; delayed cleanup actions, and failed to develop pollution prevention as the overarching and dominant strategy.[42]

Emphasizing the need for community involvement in environmental decision making, environmental justice activists charged the mainstream-professional movement with abetting government agencies and industry in excluding local communities from participation in environmental protection efforts and promoting disproportionate environmental impacts in lower-income communities and communities of color. They also railed against what Robert Bullard calls "environmental blackmail": the siting of polluting facilities in lower-income communities of color in exchange for jobs and other much-needed economic benefits.

Social science studies from as early as 1967 have confirmed disparities in environmental protection and health according to race and income, culminating in the landmark 1987 report by the United Church of Christ, *Toxic Wastes and Race in the United States,* which found that people of color were twice as likely as whites to live in communities with a commercial hazardous waste facility and three times as likely to live in a community with multiple facilities or a large hazardous waste facility.[43] In 1992, a *National Law Journal* report found that EPA is 20 percent slower to register hazardous wastes sites in minority communities on the Superfund National Priorities List than in white communities. The report also noted that EPA performs a lesser degree of remediation on these sites and levies lower fines on polluters than in white communities.[44]

In 1990, a group of environmental justice leaders representing a host of civil rights organizations from around the country wrote a letter to the heads of eight Group of Ten organizations charging them with racism in their hiring practices.[45] At that time, the Sierra Club employed no blacks or Asians and only one Hispanic among its 250-member staff. The National Audubon Society could claim only three blacks among its 350 employees. Environmental justice activists harshly criticized what many of the mainstream-professional movement's leaders already tacitly acknowledged: that environmentalism was elitist and undemocratic. As the Sierra Club president Michael Fischer stated in 1992, describing the social attitudes of the club's membership and, by extension, of the public interest environmental establishment as a whole, "National Sierra Club members just couldn't imagine themselves sitting down and talking with blue-collar or lower-income people."[46] Moreover, environmental justice leaders pointed to the parochialism of the public interest environmental groups' program and policy agenda, its romantic-progressive slant that privileged protection of wilderness areas and endangered species while marginalizing or ignoring altogether the issues most important to people of color and lower-income Americans, such as urban environmental health, toxic waste, and environmental hazards on the job.

As a result of the caustic critique of mainstream-professional environmentalism brought by environmental justice activists, both public interest environmental organizations and EPA initiated changes aimed at diversifying their programs and personnel and confronting the activists' claims of disproportionate environmental harms and unequal environmental protection in communities of color. These actions forced mainstream-professional environmentalists to take a hard look, often for the first time, at the very culture of mainstream-professional environmentalism and the principles on which it is based.

Yet the transformation of mainstream-professional environmentalism has proved a difficult task. African American environmentalist Vernice Miller, a noted environmental justice activist who cut her teeth battling air pollution and other environmental hazards in New York's West Harlem throughout the 1980s, was recruited by NRDC in 1993 to help direct the organization's new environmental justice initiatives.

"Like many other national environmental groups, NRDC has come a long way from where it was in the early 90s and before," Miller remarks, "but the elitism and parochialism of traditional environmentalism is deeply embedded in the culture of these organizations, and it will take some time, perhaps decades, before public interest environmentalists truly speak for and represent all of the public, especially people of color." She adds, "Both the ideology and methodology of mainstream groups like NRDC often undermine genuinely democratic environmental activism because, focused on wilderness on the one hand and law and policy on the other, they fail to resonate with most working-class and lower-income Americans."[47]

NRDC's recent effort to help spur environmentally sustainable development in the South Bronx highlights the challenge that mainstream-professional organizations face in working with and in minority neighborhoods and in reorienting their efforts away from traditional litigation and advocacy strategies to local, community-based initiatives. In 1994, NRDC scientist Allen Hershkowitz approached a South Bronx community development corporation, Banana Kelly Community Improvement Association, with an idea to build a state-of-the-art paper recycling mill on a contaminated, abandoned 30-acre rail yard in the Mott Haven/Port Morris section. The force of the proposal, known as the Bronx Community Paper Company, came from the fact that not only would the facility provide 600 permanent jobs to one of New York City's most economically distressed areas (with unemployment rates as high as 75 percent), but the project would be a model of green business development: a low-emission, energy- and water-efficient recycling facility that would help deal with the 10,000 tons of waste paper produced every day by residents and businesses in New York City, thus sparing the region's overtaxed landfills. With Banana Kelly as a partner, NRDC moved forward with the project, acting as a developer and consultant to two paper companies, S. D. Warren and the MoDo Paper Company. They even retained Maya Lin, the noted architect of the National Vietnam Memorial in Washington, D.C., to design the facility.

However, another South Bronx group, the South Bronx Clean Air Coalition, objected to the paper plant proposal on the grounds that pollution from the facility would exacerbate the area's already poor air quality conditions. The group waged a rancorous battle against both NRDC and Banana Kelly, claiming they failed to reach out to South Bronx residents and were forcing a polluting facility down the throats of an environmentally beleaguered community.

Miller, who played the role of mediator in the struggle, felt torn between NRDC's good intentions and promising proposal and the powerful environmental justice claims of the Clean Air Coalition that resonated deeply with her. After several years of litigation trying to halt the project, a New York appeals court ruled in 1997 that NRDC and Banana Kelly could proceed with the project. The victory, however, came at a cost to NRDC's image as a genuine proponent of environmental justice and merely confirmed the extent of the long-standing clash between mainstream-professional and grass-roots environmentalists.

Meanwhile, EPA has implemented a number of environmental justice programs and recruited and promoted many minority staff, especially among the professional ranks, since the early 1990s. Lois Adams, regional coordinator and team leader of the Urban Environmental Program in EPA's New England office, and an African American, has seen both success and struggle in EPA's efforts to promote environmental justice and to reform its mainstream-professional culture. According to Adams,

> The regions have accepted the responsibility of identifying what the critical environmental justice issues are, like childhood lead poisoning, and the best way to address them. EPA's success has come from its limited focus on small "learning laboratories" and opportunities for demonstrable improvements in the environmental justice area, to dispel myths that poor people and people of color don't care about environmental issues and are not prepared to deal with them. As a science- and engineering-based agency, supporting community-based efforts is not an easy task and cannot be measured by traditional means. Our role is really to help empower local efforts, so our success comes when others do something that EPA supports; we are a technical advisor, champion, facilitator, and funder. This is not a typical EPA role.[48]

EPA's greatest challenge, Adams explains, stems from its legal-scientific culture, which has been slow to embrace and institutionalize environmental justice strategies such as community planning and multi-stakeholder decision making. Reaching out to lower-income and minority communities and empowering them to solve environmental problems, Adams laments, has yet to become standard operating procedure at EPA.

Following the issuance of President Bill Clinton's executive order in February 1994 directing all federal agencies to ensure that their programs do not result in unfair environmental burdens being placed on minority communities, EPA stepped up its enforcement activities in the area of civil rights, especially Title VI of the 1964 Civil Rights Act. EPA's Office of Civil Rights considers complaints of discrimination filed under Title VI, which prohibits state agencies receiving federal funds from taking actions such as issuing environmental permits that unfairly burden minority communities. In early 1998, EPA issued guidance to the Office of Civil Rights providing that even when a pollution permit satisfies existing environmental laws, it might still be illegal under the civil rights law if it contributes to a pattern of disproportionate pollution in a community of color.

This policy has met with considerable resistance from state environmental agencies and business groups claiming the guidance will unduly burden their efforts to bring job opportunities and brownfields redevelopment to minority communities. "It runs contrary to Federal programs designed to bring jobs and cleanup to low-income and minority communities," lamented William Kovacs, vice president for environment at the United States Chamber of Commerce.[49] Framing the issue as a choice between economic development and environmental protection, critics of EPA's Title VI policy raise the specter of environmental blackmail that Robert Bullard and other environmental justice leaders have warned against. Nevertheless, the question remains whether tougher rules for permitting facilities in minority communities mean economic suicide. Must these communities choose between environmental degradation and economic growth?

As NRDC's experience in the South Bronx shows, environmental protection and economic development can go hand in hand, thus meeting the needs of many lower-income and minority neighborhoods. But such strategies demand considerable effort on the part of mainstream-professional environmentalists to reach out to local community groups and earn their trust. As the mainstream-professional movement attempts to recreate itself in the light of the criticism from environmental justice and other grass-roots activists, it will have to engage local communities and their issues directly. As the history of American environmentalism demonstrates, economic development and grass-roots organizing have never been at the core of the movement's agenda. To the contrary, economic development has traditionally been conceived by environmentalists as anathema to environmental protection. Comprising largely the affluent and privileged, most environmentalists have never had to worry about finding a job or the problem of economic disinvestment in the communities where they live and work. As a professional matter, environmentalists have followed John Muir in their pointed attack on industrialism as the scourge of environmental protection while ignoring the concerns of working-class Americans.

Moreover, the evolution of American environmentalism, from the romantic-progressive phase to the mainstream-professional era, reveals a steady and deliberate turn away from locally oriented activism to centralized, professional advocacy. Local communities and constituencies have been relegated to the back burner by both the public interest environmental establishment and the environmental law and policy system itself.

The Great Divide Between Democracy and Environmentalism "If the overriding objective of environmental activism is protection of the entire environment," Mark Dowie writes, "the traditional environmental movement was no more than half a movement. Limited from the start, it was almost obsessively oriented toward wilderness, public land, and natural resources conservation."[50] Left out of the movement have been the people themselves and the environmental issues that, quite literally, hit home—local issues like lack of open space, brownfields, asthma brought on by air pollution, and other environmental problems endemic to many American communities.

Picking up on this theme, William Cronon decries American environmentalists' habit of "idealizing a distant wilderness" at the expense of the local, the everyday. Such idealizing, he explains,

> means not idealizing the environment in which we actually live, the landscape that for better or worse we call home. Most of our most serious problems start right here, at home, and if we are to solve these problems, we need an environmental

ethic that will tell us as much about *using* nature as about *not* using it. . . . [Respect for nature] means never imagining that we can flee into a mythical wilderness to escape history and the obligation to take responsibility for our own action that history inescapably entails.[51]

In effect, the proud tradition of American environmentalism has mirrored the same democratic deficits that have afflicted the society as a whole. Bureaucratic, centralized, and technical, modern mainstream-professional environmentalism has largely ignored local communities and the civic networks necessary to sustain them. Notwithstanding the rich tradition of grass-roots environmentalism dating back to the neighborhood organizing of Jane Addams's Hull-House and still extant today in many communities across the country, mainstream-professional environmentalists have not used their considerable power to foster community-based environmental efforts and have even sometimes helped to undermine them, as in the case of federal environmental standards like the National Ambient Air Quality Standards that ignore local air pollution hot spots, or the disproportionate impact of polluting facilities in minority communities for too long sanctioned by environmental regulators and public interest environmentalists alike. By focusing their problem-solving strategies on the courts and Congress, and on issues like protection of wilderness areas, parks, and endangered species, mainstream-professional environmentalists have generally avoided local environmental issues and discounted the value of local civic networks in addressing environmental harms. Consequently, they have indirectly countenanced the continued environmental degradation of local communities.

Meanwhile, these self-same strategies have proved unreliable in generating lasting environmental protection outcomes. Citizen suits, for example, the bread and butter of many mainstream public interest environmental campaigns, have lost much of their power. Rulings by the Supreme Court and several lower federal courts have dramatically curtailed the ability of citizens and their environmental organizations to sue violators of pollution laws.

Increasingly, conservative federal judges are invalidating congressional provisions granting citizens the right to enforce environmental laws by raising the threshold that plaintiffs must meet in showing real harm from alleged violations.[52] Essentially the courts have held that if polluters stop polluting after a citizen suit has been filed, plaintiffs cannot show they are harmed, nor can their lawyers recover attorneys' fees, which traditionally have financed these suits. As one environmental lawyer has remarked, many federal judges "are trying to close the courthouse doors. It means fewer and fewer cases will be brought, and you will have more and more environmental problems because the government cannot—and in some instances will not—pick up the slack."[53]

In another recent example, the federal courts invalidated new air quality rules promulgated by the EPA to regulate fine-particulate matter. Tiny particles of toxic soot less than 2.5 microns in diameter, so-called fines are emitted by common sources like automobiles and power plants. Because of their size, they stay trapped in the lungs and respiratory system, making them especially hazardous to human health. Responding to the widespread call for tighter controls on particulate pollution resulting largely from the urban asthma epidemic, EPA developed the rules in 1997, only to see them second-guessed by a federal court in 1999. The court held that EPA had not sufficiently shown it had the power to require emissions standards for a new type of pollutant and, much like the federal courts' reaction to citizen suits, fundamentally questioned the power of environmental laws to hold polluters accountable. The fate of the fine-particulate regulations, and of the countless thousands who suffer from asthma, remains uncertain. Such is the nature of environmental regulation.

With its emphasis on legal and technical solutions, mainstream-professional environmentalism has failed to encourage active political and civic participation. The legal and technical nature of mainstream-professional environmentalism has given it the appearance of being almost apolitical—beyond politics and in the realm of neutral expertise. Viewed as a high-powered elite with access to government and industry leaders, mainstream-professional environmentalists have inadvertently disempowered local constituencies, who are often left feeling they are not heard in determining environmental outcomes and that the issues that matter to them are generally not of concern to mainstream-professional environmentalists. With its direct mail machinery, centralized structure, and top-down decision making, mainstream-professional environmentalism has cultivated a largely passive constituency and in the process has stripped itself of the ability to activate and inspire robust political participation and civic engagement, the very forces that can hold decision makers accountable, prevent environmental harms, and institute local and regional environmental strategies like smart growth and brownfields redevelopment.

This kind of environmentalism has relied on legal and advocacy tools often at the expense of other techniques, such as community and regional planning, which both engages local stakeholders and establishes meaningful environmental and social goals. Planners and their expansive toolbox of ideas and practices aimed at building livable, health communities comprise a separate tradition from mainstream-professional environmentalism. In the light of the traditional disjunction between environmental law and policy on the one hand and land use and planning on the other, the failure of environmentalism to embrace proactive, planning-oriented strategies is understandable, though unfortunate. The ideas of visionary planners, from Frederick Law Olmsted, Lewis Mumford, and Jane Jacobs, to Benton MacKaye, Ian MacHarg, and Robert Yaro, have remained outside the sphere of mainstream-professional environmental advocacy, resulting in a largely after-the-fact approach to environmental problem-solving.

Moreover, without an engaged constituency behind them, mainstream-professional environmentalists have been at a disadvantage in dealing with the influence of corporations and other private actors on the design and enforcement of environmental laws. As new initiatives like Project XL and ERP demonstrate, even the most well-intentioned policies can falter in the absence of the full participation of community stakeholders. Mainstream-professional environmentalists have not prepared these stakeholders to participate in initiatives like Project XL or ERP and have failed to marshal the broad-based political power necessary to level the playing field of environmental law and policy.

The elitism and homogeneity of traditional environmentalism is further evidence of the movement's democratic deficits. Only recently bothering to reach out across racial or economic lines, mainstream-professional environmentalism has alienated racial minorities and the working class, who traditionally have not identified with environmentalists. Moreover, notwithstanding the progress of the past several decades, environmental harms have not let up in lower-income and minority communities, revealing a gap in mainstream-professional environmentalism's advocacy agenda or, worse, confirming the success of the environmental law and policy system. As environmental justice attorney Luke Cole writes,

Environmental laws are not designed by or for poor people. The theory and ideology behind environmental laws ignores the systemic genesis of pollution. Environmental statutes actually legitimate the pollution of low-income neighborhoods. Further, those with political and economic power have used environmental laws in ways which have resulted in poor people bearing a disproportionate share of environmental hazards. . . .

Mainstream environmentalists see pollution as the *failure* of government and industry—if the environmentalists could only shape up the few bad apples, our environment would be protected. But grassroots activists come to view pollution as the *success* of government and industry, success at industry's primary objective: maximizing profits by externalizing environmental costs. Pollution of our air, land, and water that is literally killing people is often not in violation of environmental laws.[54]

Because environmental laws do not prevent pollution but merely control it, and decisions about the distribution of environmental benefits and burdens such as parks and polluting facilities are naturally a function of the relative political power of communities, it is no accident that environmental hazards persist, often following the path of least resistance to lower-income and minority neighborhoods. As Luke Cole declares, "It is because the law works for white middle-class communities that it does not work for the poor, or for people of color."[55] Lacking the political, medical, economic, and legal resources more affluent communities possess, and facing environmental hazards of all kinds, lower-income and minority neighborhoods are at the greatest risk of harm. The physical conditions in these neighborhoods thus represent some of the most serious environmental problems of our time. Yet the mainstream-professional movement has just begun to take notice.

Further, because mainstream-professional environmentalism has never articulated a compelling vision of economic development commensurate with its emphasis on the protection of natural resources and wilderness, notwithstanding Gifford Pinchot's rhetoric about "sustainable yield," economically distressed communities have failed to gravitate toward the movement while industry has gone about its business in the usual fashion. Despite the importance of economic investment and employment opportunities to overall community health, including environmental protection and the availability of environmentally sustainable production methods, these issues have not risen to prominence among mainstream-professional environmentalists. Similarly, the movement has failed to address the persistent segregation of communities along racial lines, which has resulted in the continuing development of suburban and rural areas, with the associated environmental costs.

Lacking a democratic vision and an active, diverse constituency, mainstream-professional environmentalists have also been unable to stand up to the pervasive forces of privatization and global capitalism. As EPA and state agencies creatively attempt to improve environmental protection efforts through regulatory reinvention, they risk empowering corporations and other polluters at the expense of local communities, which have yet to participate meaningfully in policy discussions and inevitably bear the brunt of industrial pollution as workers and plant neighbors. Because with privatization invariably comes the loss of accountability, EPA and its state counterparts must ensure that the public is also involved in entrusting the private sector with environmental leadership. So far, mainstream-professional environmental organizations have proved unwilling or unable to galvanize an active public constituency that can level the playing field of regulatory reinvention.

Further, mainstream-professional environmentalists have not argued vigorously enough for increased public investment in environmental protection. Public goods demand public investment, which helps to compensate for capitalism's short-term, profit-centered motives. The market has traditionally proved resistant to correcting its environmental failures, or externalities, like brownfield sites and air pollution from cars and trucks; it has failed to internalize environmental costs and has lacked the will and soul to achieve lasting environmental results. Public investment is thus needed to fill the myriad gaps left by the market and manifest in widespread environmental degradation. Public investment can also help encourage private investment in environmental improvements like cleanup of contaminated sites and implementation of environmentally sound management and production practices.

Moreover, environmental assets like mass transportation, parks, and tree-lined walkways are paid for by public funds, and they require ongoing public investment for their maintenance and upgrade. Such assets constitute a significant part of the country's public spaces—the physical infrastructure that allows people to come together, associate face to face, and engage in civic activities. Without them, our communities possess none of the physical resources that allow civic life to be expressed. In essence, environmental assets are the enabling mechanisms for civic culture.

But public investment also entails citizens' taking responsibility for their own actions as producers, consumers, and voters. Privatization is not only a matter of the ascendancy of corporations; it is also about the abrogation by individual citizens of their obligations as public actors. *Public investment* is another name for *public consciousness*—for an awareness that each of us is a member of a larger community that rises or falls based on our commitment to the common good. Thus, individuals, whether in their capacity as corporate executives or consumers, need to exercise their public will in the decisions they make every day.

Making Environmentalism More Civic: Toward New Approaches to Environmentalism and Democracy

In the light of the persistence and pervasiveness of environmental degradation across America, and the inability of local communities as well as environmentalists to grapple with it, there is both an urgent need and a unique opportunity to retool and realign citizens and environmentalists alike so they are better equipped to improve the environmental and social conditions of their communities. Americans have come to realize that despite the economic prosperity and social progress the nation has achieved over the last several decades, something profound is missing. Most Americans have lost touch not only with their neighbors, but the physical places where they live and work—their environment. In the course of an ordinary day, week, or month, many of us have little direct involvement in the civic life of our communities, nor do we enjoy ready access to a safe, quality environment. The two are causally connected. As a public good, a healthy physical environment demands informed, active public participation in local decision making to ensure that the private sector, government, and even one's own neighbors do not undermine long-term environmental gains in their pursuit of short-term, narrow ends.

All environmental harms are local in origin, though their effects may spread great distances. In thinking about environmental degradation, we tend to get lost in abstractions like global climate disruption, or even in the minutiae, such as pollution measured in parts per billion. Competing statistics can confuse us. One day we read that the air is getting cleaner or that a certain endangered species is making a strong comeback; the next day we hear reports that water pollution continues to be a major public health threat or that remote rural areas are being developed at unprecedented rates. How do we reconcile this information? How do we make sense of the endless litany of statistics and figures that often seem at odds?

The answer is as simple as looking out the window. The true test of environmental quality is the environmental conditions on the ground, in the trenches of local communities across the nation. What do you see when you peer out the window—from your bedroom, office, or automobile? For many, the sight is as unpleasant as it is unsettling, at best offering a meager experience of nature and place and at worst presenting real and immediate health threats. Ironically, it is these same conditions that we have tended most to neglect in our environmental protection efforts. Notwithstanding the local nature of environmental harms, in terms of both their genesis and consequences, traditional environmentalism has focused on places where very few of us actually live and work, such as wilderness and national parks, while overlooking densely populated areas like cities and suburbs. It is for this reason that William Cronon warns, "Wilderness poses a serious threat to responsible environmentalism at the end of the twentieth century."[56] We must move beyond fetishizing the sublime and wild, he urges, and instead embrace the humble places most of us call home, bringing the powerful lessons wilderness teaches into the more quotidian reality of our day-to-day lives.

Moreover, our environmental regulatory system seems to have lost the forest for the trees. With its focus on large, stationary point sources such as power plants and sewage treatment facilities, its uniform standards approach, and its preference for control over prevention, environmental laws have countenanced pollution from small, dispersed sources like automobiles and auto body shops, ignored local land use decisions and regional planning, failed to address pollution hot spots, and allowed for the continued use of toxic materials, all of which directly and adversely affect local communities.

"The future of American environmentalism . . . remains outside the Beltway," Mark Dowie declares, "in the hands of a new civil authority forming in and around thousands of watersheds, forests, factories, and communities scattered throughout the country."[57] It is this civil authority, or civic environmentalism, that can bring us back to the places where we live, work, and play every day, and in the process help us reclaim those places, both physically and socially. Civic environmentalism forces us to reckon with our own communities and ourselves as the citizens, corporations, and governments who comprise them. We have met the enemy, the cartoon sage Pogo admonished, and the enemy is us.

Today's environmental problems demand local solutions, crafted and administered by the diverse stakeholders that constitute our communities. Issues that were once considered the sole province of state and federal government, like the environment, criminal justice, and education, have fallen into the laps of local communities, owing in large part to the prevailing political climate of decentralization and downsizing. In this shift from top to bottom is an opportunity to rebuild our communities, from the ground up.

The Core Concepts of Civic Environmentalism

Civic environmentalism entails a set of core concepts that embraces civic action and community planning on the part of a diverse group of stakeholders aimed at promoting both environmental protection and democratic renewal: participatory process, community and regional planning, environmental education, industrial ecology, environmental justice, and place.

Civic environmental projects reflect all of these core concepts to varying degrees. In any effort, one concept might have more prominence than another, but at bottom all civic environmental efforts entail some aspect of each, however implicit or latent. What ultimately defines civic environmentalism and distinguishes it from other forms of social action is the explicit link between environmental problem solving and the goal of community building. Civic environmentalism is fundamentally about ensuring the quality and sustainability of our communities, economically, socially, and environmentally.

Participatory Process Meaningful, informed participation in the decision-making procedures that affect the quality of people's lives is at the root of civic democratic practice and therefore civic environmentalism. As Cornel West claims, the true test of American democracy is the ability of ordinary people to participate in the decision making of institutions that fundamentally regulate their lives.[58] Civic environmentalism provides for the regular, practical participation of all citizens in environmental decisions so that environmental outcomes are the shared function of the many. This accords with Benjamin Barber's notion that strong democracy requires that all citizens participate in decision making at least some of the time.

Participation entails more than simply showing up. It demands recognition, equitable treatment, and deliberative responsibility.[59] Recognition means having decision-making procedures that convey a communal acknowledgment of equal individual worth, regardless of income, age, gender, race, ethnicity, or geography; equitable treatment frames political or public questions in a way likely to elicit outcomes that satisfy everyone's interests fairly; deliberative responsibility demands informed, open, and responsible public deliberation. These features of participation ensure that decisions are justifiable to each person who comes under their sway, regardless of her point of view or social and historical circumstances.

In addition, participation in civic environmentalism requires face-to-face interaction among diverse stakeholders, including corporate executives, developers, residents, and government officials. As Michael Briand, director of the Colorado-based Community Self-leadership Project, describes, "In the absence of face-to-face interaction there is a tendency to see political opponents as distasteful caricatures. . . . Face-to-face exchange enables people to develop a more complex, more human, more realistic picture of their fellow citizens. . . . It encourages people to live up to commitments they make in the course of reaching a public decision, making citizens more accountable, and thus also more reliable partners."[60] Such interaction also breeds and fosters a genuine sense of community, which in turn inspires more participation, creating a positive feedback loop. The process of working together to solve community problems is a powerful community-building exercise; achieving visible results profoundly compounds the sense of possibility.

Further, participation in the civic environmental model validates the expertise of not only professionals but ordinary people. It embraces a bottom-up approach to problem solving, empowering people to work together, initiate action, experiment, and learn facts. Borrowing from John Dewey's pragmatism, civic environmentalism holds that ordinary people, with the help of experts, have the opportunity to act for themselves in attempting to solve social problems. As elites, experts are trained to tell others how to solve their problems, but this top-down approach is disempowering and undemocratic. For Dewey, democracy happens when all people have the opportunity to develop and use their capacities to the fullest extent possible; it fails when the privileged "shut out some from the conditions which direct and evoke their capacities."[61]

Perhaps there is no other issue more fundamental to the quality of people's lives than environmental degradation, whose consequences can be a matter of life or death for those directly harmed. The scope and significance of environmental issues dictate that environmental decision making be as inclusive and participatory as possible. Moreover, the very nature of environmental harms compels the participation of a diverse set of stakeholders. Pollution problems defy all borders—racial, cultural, political, and geographic; even in its most localized form, environmental degradation is never the concern of just a few individuals, but reaches across streets and districts to become an entire neighborhood's problem. As well, in the light of the technical and scientific nature of environmental problems and the complex regulatory framework established to deal with them, information sharing and education among a diverse group of stakeholders are essential to ensure effective multistakeholder participation. Experts, consultants, government officials, and citizens must be able to discuss and exchange ideas and data efficiently and in the open, with careful attention paid to the interests of laypeople, who may lack the technical expertise of environmental professionals yet often possess important anecdotal information.

Civic environmentalism thus moves beyond the professional model of traditional environmental problem solving by engaging a participatory, multi-stakeholder process. This process allows for information to be shared and understood among a diverse group and around a particular problem or set of problems. Moreover, the process of multi-stakeholder participation helps develop and strengthen the social capital and civic infrastructure necessary to carry out effective environmental programs and establishes a working model for decision making that can be replicated across issues and problems. Active participation enables citizens to monitor problems, identify opportunities for improvement, support local initiatives and development activities on the part of government and the private sector, and help create comprehensive, innovative approaches to community revitalization efforts as a whole. At the same time, multi-stakeholder participation increases a sense of community; improves relations and accountability among government, corporate, and citizen stakeholders; and induces corporations, government agencies, and citizens to act responsibly.

Community and Regional Planning Effective community and regional planning starts with the notion that local communities want to ensure that future generations will inherit a place to live that is physically and socially healthy and vital. This is the essence of the idea of sustainability. It flows from and is a consequence of active, multi-stakeholder participation. That is, civic environmental planning assumes that some measure of civic infrastructure is in place to carry out a meaningful, comprehensive planning process. Without a stable civic infrastructure, communities cannot engage in genuine planning projects, whether on a project or community-wide basis. Commenting on a major, multiyear planning process for the development of Boston's waterfront, for example, one participant remarked that local stakeholder groups need "structures to get us to 'yes' more quickly, without wasting time and resources. . . . We need a larger conversation and better civic capabilities."[62] Participation and planning are thus mutually dependent practices that help develop a strong sense of community and common purpose.

Civic environmental planning is the response to the unplanned, uncoordinated growth that has characterized most of the nation's physical development since the arrival of the first European settlers in the seventeenth century, resulting in pervasive negative environmental and social effects such as blighted, abandoned urban neighborhoods, gated suburban communities, and loss of wildlife habitat. It borrows from the rich tradition of community and regional planning exemplified by the work of Frederick Law Olmsted, Benton MacKaye, Lewis Mumford, Jane Jacobs, Kevin Lynch, Ian McHarg, Robert Yaro, Ann Whister Spirn, and many other planners who have often been our most articulate, yet least heeded, spokespeople for healthy, sustainable communities. Dominated by lawyers and policymakers, environmental problem solving has tended to rely on ex post legal and administrative solutions rather than proactive, planning-based solutions. Like ordinary citizens, planners have frequently been left out of environmental programs and projects.

Civic environmental planning involves the process of moving beyond piecemeal problem solving to understanding the systemic issues that underlie those problems and designing comprehensive, long-term solutions. In fashioning those solutions, it looks at not only the problems besetting a community but, more important, the community's assets—that is, what is working for the community. Further, civic environmentalism allows stakeholders to envision their community five, ten, or twenty years down the road and to take stock of the resources necessary to achieve that vision. Taking the long view also forces stakeholders to account for consequences that would otherwise be externalized or ignored, shifted to some other place or time.

Further, civic environmental planning provides an opportunity to explore the connections among seemingly unrelated social policy issues, such as the intersection between criminal activity and environmental degradation (for example, illegal dumping of hazardous waste on inner-city vacant lots), or urban disinvestment and loss of rural open space, and to identify key community objectives in the light of local, state, national, and even international trends.

Civic environmental planning also bridges the gap between environmental regulation and local land use and development issues, embracing new strategies for sustainable development, or preventing development altogether when necessary. Such strategies take a regional approach, linking the common concerns of urban, suburban, and regional constituencies in public goods like open space protection and environmental quality. The focus is on coordinating regional land use and development, and devising systems of governance bigger than local but smaller than state and federal governments that are matched to the scale of regional problems. Whereas traditional environmental laws do not directly deal with local and regional land use practices or sprawl, land use planning and smart growth efforts allow communities to go beyond the narrow jurisdiction of environmental law to devise measures aimed at achieving common goals like open space protection and public access to waterfronts and parks, while encouraging investment in urban centers and mass transit. Within constitutional limits protecting the interests of private property owners, communities can adopt zoning ordinances and other laws designed to control development. As well, they can establish local taxes, such as transfer taxes on real estate transactions or property taxes, that generate funds for purchasing land for open space and parks. Other techniques, such as conservation easements and purchase of development rights, are also available. Under the planning model, regulation is a means to an end, not an end in itself, and is just one of many smart growth tools.

Moreover, unlike the pollutant-by-pollutant, media-based approach of environmental law, civic environmental planning allows stakeholders to view the local environment as a true ecosystem, composed of intricately interwoven parts. This ecosystem approach, exemplified by the growing trend toward local watershed initiatives, in which communities within a common watershed monitor and control activities as they affect the ecosystem as a whole, accounts for the redistribution of pollution across media and neighborhoods resulting from pollution controls, and encourages pollution-prevention measures.

To create this dynamic planning framework, civic environmentalism embraces a systems approach. Communities comprise a variety of systems—social, environmental, economic, and political—which depend on accurate information to function well. Without good information, communities cannot effectively manage those systems or implement long-term plans. Accordingly, communities need feedback mechanisms to generate information about a system that might otherwise be too complex to understand.

An innovative feedback mechanism for community planning has been developed by the public policy group Redefining Progress and its Community Indicators Network. This approach uses community indicators, designed by community stakeholders, that function like instrument panels, providing citizens, government agencies, and businesses with information about past trends and existing realities, giving them guidance for action and allowing them to prioritize issues in the light of limited human or financial resources. Community indicators enable citizens to measure their progress toward their vision and ensure that their efforts result in positive changes.

Traditional planning techniques tend to focus on indicators like job growth and housing starts, interpreting increases as evidence of community health. However, because such increases might amount to environmental harms, such as loss of open space and air pollution, these indicators can be misleading. Community indicators focus on environmental, economic, and social conditions in an attempt to present a more complete and accurate picture of community life. They allow stakeholders to integrate these different concerns and to design systems-oriented strategies around them.

Indicators are "small bits of information that reflect the status of larger systems. . . . When we can't see the condition of something directly and in its entirety—whether it's a car, a person, an educational system, or a whole community—we need indicators to make those conditions visible. Indicators can't tell us everything, but they can tell us enough to make good decisions possible."[63] Community indicators thus enable diverse stakeholders to make informed decisions and to be accountable for the results. Moreover, as Redefining Progress notes, "they are powerful enough leverage points within systems that they can make change happen almost on their own, even without any recourse to new regulations, programs or policies."

For example, when the federal Toxics Use Reduction Act went into effect in the late 1980s, requiring users of toxic pollutants to list those pollutants for public review in the Toxic Release Inventory, the inventory itself became an indicator. The existence of this information alone was sufficient to reduce toxic emissions, as companies cut back on their use of toxic materials so as to avoid negative publicity. Other environmental indicators include the number of fish species returning to spawn in local rivers; the amount of green space as a percentage of overall municipal land area; the amount of impervious surface area in a city; the number of beaches open for swimming; the amount of waste per capita per year; the incidence of environmental health problems like asthma; the number of green businesses in the area; and the number of alternative-fuel vehicles in operation in a city or town.

Civic environmental planning looks to innovative techniques like those developed by Redefining Progress. It provides the occasion for leveraging civic resources in the service of concrete environmental goals linked to an overall community vision. Whether in a formal or informal setting, it is a proactive, asset-oriented approach that recognizes the interdependence of business, government, nonprofit organizations, and citizens and the various systems that comprise their community.

Environmental Education Education is a central component of civic environmentalism because it helps enforce the notion that environmental and social conditions are mutually reinforcing and that local communities possess the power to change their circumstances. It is also a precursor to both participation and planning in that it enables stakeholders to make reasoned and informed judgments about environmental issues. In the civic environmental model, environmental education entails informing both producers and consumers about the environmental and social costs associated with economic activities in the hope they will change their practices and develop a strong environmental and civic consciousness. For instance, a local company that uses hazardous materials should be educated about safer alternatives and cost-effective ways of using them, and should be publicly supported in making the switch. Consumers should know about the environmental consequences of their actions, such as air pollution and global warming stemming from vehicle miles traveled in cars and sport utility vehicles, or habitat destruction as a result of suburban development, so they can make informed, environmentally responsible choices. Citizens, government agencies, and businesses also need to learn about the disproportionate environmental harms that tend to fall on lower-income and minority residents. Such awareness can lead to more equitable decisions about the allocation of environmental benefits and burdens and to policies that emphasize pollution prevention so that no community or population has to suffer undue environmental hazards.

Every American citizen, corporation, and government agency should be expected to possess a bare minimum of environmental literacy, much as they are expected to have general knowledge about other socially relevant issues like the way laws are passed, or the history of the civil rights movement, or the significance of the Dow Jones Industrial Average. Our schools, media, and civic institutions must ensure that the relationship between environmental harms and social and economic change is commonly understood and that the environmental consequences of both private actions (driving a gas-guzzling automobile) and public policy (investing tax dollars in roads over mass transit) are routinely identified and discussed as a matter of course.

Moreover, in the light of the scientific nature of many environmental problems, technical data and information must be disseminated to stakeholders in an intelligible, meaningful way to allow them to make sound decisions. For example, in setting priorities for spending on environmental protection initiatives, local communities need to understand the comparative health risks associated with different forms of pollution, for instance, air pollution versus brownfields. In the absence of endless resources, community stakeholders have to be able to make prudent judgments about which environmental problems or opportunities to pursue based on the comparative risks posed to human health and ecosystems, while maintaining a commitment to systemwide pollution prevention.

Further, to the extent that citizens are engaged in actual environmental projects, such as monitoring air emissions from local industrial facilities or sampling water quality from nearby streams, they need the technical knowledge required to perform these tasks. With just a little education, citizens can become volunteer experts, reducing the cost while increasing the effectiveness of environmental protection programs. Since the days of Jane Addams, citizen-experts have provided invaluable information about environmental problems, linking public health threats like certain cancers to environmental causes through their "barefoot epidemiology."

Environmental education also helps build social capital by creating a positive feedback loop: as local stakeholders become more knowledgable about environmental issues, they will become more interested in participating in environmental decision making and program implementation. Environmental education also develops awareness among stakeholders of the civic resources and regulatory system that promote environmental protection.

Industrial Ecology Civic environmentalism looks to industrial ecology (IE) as an innovative conceptual and practical model for integrating environmental protection measures with economic and community development initiatives. A new field of environmental and economic policy, IE has been called the practice of sustainable development and denotes the idea that the industrial processes of extraction, production, distribution, and consumption should function in roughly the same way as ecosystems, where unused waste is the exception, not the rule, environmental impacts are minimized, and a diversity of life forms is sustained over time. IE signals a shift away from end-of-pipe, media-based pollution control methods toward strategies for more comprehensive, integrated pollution prevention and planning of environmentally sound industrial and economic development.

IE incorporates tools such as full-cost accounting and design for the environment (also called "green design") for integrating environmental objectives into production processes, product design, and decisions about selection of materials and technology. IE also looks at the ways in which firms can share resources, such as water, power, and waste, resulting in closed-loop operations.[64]

Traditional pollution control methods such as smokestack scrubbers end up redistributing pollution from one medium (air) to another (soil), rather than eliminating or reusing waste. Traditional environmental regulation also has fragmented the stages of a product's life cycle. Thus, regulation focused on a factory's emissions does little to reduce environmental impacts caused when the materials used in a facility have already been produced by other suppliers, or when the product itself reaches a firm's customers, whether they are distributors, retailers, or end users. IE thus calls for a life cycle approach to industrial production—evaluating and preventing the adverse environmental impacts of a product or service from the point of extraction to consumption. It entails a systems or holistic approach to understanding the complex relation among various kinds of human activity—economic, social, political, cultural—and the environment. As the name implies, IE fuses together two concepts, industry and ecology, that have traditionally been viewed as mutually exclusive, resulting in a compelling paradigm of economic development and environmental protection. As the noted architect William A. McDonough explains, industrial ecology opens the door to the "next industrial revolution" in which pollution is prevented, not controlled, and industry functions like an ecosystem, converting waste into food.[65] McDonough and his firm, William McDonough and Partners, wrote "The Hannover Principles," summarized below, which outline the fundamental concepts of sustainable design:

1. Insisting on the rights of nature and humanity to coexist in a healthy and sustainable way.
2. Recognizing the interdependence between human action and the natural world.
3. Respecting the relationship between the community, its dwellings, industry and trade.
4. Accepting responsibility for the consequences of design on human well-being and the viability of natural systems.

5. Creating safe objects of long-term value; relying on natural energy sources.
6. Understanding the limits of design in solving all problems.
7. Seeking constant improvement in the sharing of knowledge.[66]

Stemming from the ideas of designers, architects and engineers like McDonough, Amory Lovins, John Todd, and David Orr, and not from lawyers or policymakers, IE represents a shift in the way we think about pollution and pollution control. Focused on industrial activities in the broadest terms, IE looks to the private sector for environmental leadership while forcing consideration of the full range of environmental, social, and economic issues associated with industrial activities. In this way, it is the consummate planning tool.

Moreover, IE provides environmentalists with a compelling model of economic development, enabling them to engage and promote economic development and the built environment as a legitimate environmental issue. As part of civic environmentalism, IE's value comes as much from its use as a community planning tool for pollution prevention as a technical, workplace- and production-oriented strategy. It provides not only engineers and architects, but government agencies, environmentalists, and community groups an innovative way of looking at development challenges—one that accords with environmental values and community vision.

Ecoindustrial parks (EIPs) are an example of IE in action. First developed in northern Europe, EIPs are a collection of firms located within an ecologically designed park, each of which exchanges materials and energy—for instance, converting one firm's waste into another's feedstock or raw materials. The Stonyfield Farm Yogurt Company in Londonderry, New Hampshire, is in the process of developing an EIP designed to be the most advanced in the world. Still in the design phase, the Stonyfield Farm EIP would encompass 100 acres of woodland, preserving approximately 80 percent of the park for habitat and recreation. The plans call for state-of-the-art, energy-efficient, and architecturally appropriate facilities, housing a diverse set of firms, with minimal generation of waste. For example, rooftops would be used for ecological purposes such as increasing the vegetated area of the park, generating solar energy, and converting sewage to fertilizer. In addition, the EIP would offer educational and training programs related to IE.[67]

IE projects are not confined to EIPs. They can include any economic enterprise that incorporates techniques like full-cost accounting or design for the environment. Agriculture, construction, and other commercial or public ventures can come under the IE model. In essence, any development project or economic activity can be approached using IE principles to achieve pollution prevention and other environmental benefits.

IE is the logical flip side of brownfields policy. Brownfields is not simply about cleaning up and reusing contaminated sites to spare clean sites from development and to bring needed economic development to degraded urban areas; it is also about ensuring that brownfields sites will not simply be recreated for future generations to clean up. Redevelopment must be clean and sustainable; it must not itself bequeath a toxic legacy when it comes time for a facility to be shut down or relocated. To the extent that the practice of IE minimizes or prevents the generation of hazardous waste, habitat destruction, and worker and consumer exposure to health risks, it is the necessary companion piece to brownfields policy. Indeed, IE compels this closed-loop approach to brownfields redevelopment, just as brownfields policy itself is a response to the unintended consequences of traditional hazardous waste site cleanup laws. In this sense, IE is as much about sustainable land use practices and community planning as it is about state-of-the-art production methods.

Environmental Justice Civic environmentalism also demands an awareness of the distributive aspects of environmental protection and a commitment to democratic justice. It holds that democracy works best when everyone lives, works, and plays in a safe, healthy environment and that social justice implies environmental health for all. Civic environmentalism promotes the Native American conception of the environment as a key to a healthy community. As Michael Delaney, tribal judge of the Vermont-based Abenaki of the Mazipskwik, explains, the political, social, and cultural life of Native Americans is inextricably linked to environmental health because "the environment is not something 'out there', but something deep within each of us, a part of each of us."[68]

Environmental justice emphasizes the structural conditions surrounding the democratic process and asks the question, "Are ordinary people, especially the disenfranchised, participating in the decision-making procedures of institutions that fundamentally regulate their lives?" By striving to empower individuals to participate in and take control over decisions that affect their health and environment, environmental justice is an effort to build up the civic capacities of communities and to ensure lasting environmental results at the community level.

Place The forces of development, production, and consumption can corrode the sense of place that is essential to a community's civic and environmental health. Place serves as a mirror into the soul of a community and reinforces its cultural and social conditions. Civic environmentalism promotes the notion that "a place is a piece of the whole environment that has been claimed by feelings"[69] and embraces the idea of place put forward by the environmental philosopher Mark Sagoff:

> If you want to understand what makes the economic use of environmental resources sustainable—if you want to know how places survive the vagaries of markets—then look to the relationships, cultural and political, of the people in them. Look for affection not for efficiency as the trait with which people treat their surroundings. Where family and community ties are strong, where shared memories and commitments root people to a place, they can adapt to changing conditions, and they will do so in ways that respect nature and conserve the environment.[70]

Perhaps no one has described the democratizing and unifying power of place better than the poet Gary Snyder:

> Of all the memberships we identify ourselves by (racial, ethnic, sexual, national, class, age, religious, occupational), the one that is most forgotten, and that has the greatest potential for healing, is place. We must learn to know, love, and join our place even more than we love our own ideas. People who can agree that they share a commitment to the landscape—even if they are otherwise locked in struggle with each other—have at least one deep thing to share.[71]

Civic environmental strategies therefore attempt to develop and reinforce a sense and experience of place to help ground citizens to their communities and bolster their sense of a shared destiny. Place is the physical, social, and emotional space that nurtures us as individuals and members of a community and is an animating force for civic engagement.

Civic Environmentalism as the Emergent Paradigm

Civic environmentalism represents a synthesis of environmental and social goals framed by democratic principles and a commitment to sustainable economic development. It is as much about process as it is about outcomes, viewing the two as inseparable and mutually reinforcing. Most important, civic environmentalism embraces the idea that a diverse group of stakeholders, from professionals to government officials to ordinary citizens, can effect long-term environmental and social change in their communities with the proper mix of collective will and social capital.

Further, civic environmentalism holds that as a society, we possess most of the ideas and technologies that can significantly advance our efforts to achieve socially and environmentally healthy communities. We even appear to have the will, with poll after poll showing upwards of 75 percent of the American public in favor of strong environmental protection programs (without regard to cost), and most Americans desiring more civic involvement and a greater sense of community. Yet we lack the coordinating mechanisms and civic infrastructure necessary to put our ideas, technologies, and will into effect. There are too few working models of sustainable communities to inspire and guide us.

Nevertheless, they do exist, even if only as works-in-progress. Across the nation, local initiatives now underway demonstrate civic environmentalism in action. Innovative experiments, each models or approximates many or all of the principles that define civic environmentalism. They serve as beacons of a new environmentalism capable of meeting the challenges of the twenty-first century and at last fulfilling environmentalism's promise as American democracy's great symbol.

In the following chapters, I describe four case studies in civic environmentalism, in Boston, Massachusetts; suburban New Jersey; Oakland, California; and Routt County, Colorado. These communities are part of an emerging movement, rising up from our cities, suburbs, and rural areas, that is simultaneously protecting the environment and improving civic life.

Borrowing from a long tradition of community activism and environmental stewardship, these case studies, though still incomplete, are object lessons in turning vision into reality with environmental principles and civic commitment. They are an important part of the democratic turn in environmental action, gradually but steadily rounding the corner between what has come before and what is yet to be.

4

Urban Agriculture in Boston's Dudley Neighborhood: A Modern Twist on Jefferson's Dream

Our urban agriculture initiative is not only about helping improve the neighborhood's physical and economic condition, but generating a kind of civic alchemy that is part of creating our urban village.
—Greg Watson, interview

The farm not only grows food, it nourishes human character, is a challenge and a teacher, a source of insight and values.
—Wendell Berry, *The Unsettling of America*

From Colonial Farming Village to Postindustrial Wasteland: A Brief History of the Dudley Area through the 1980s

It all started with the land. Located less than two miles from the heart of downtown Boston, Massachusetts, the Dudley neighborhood straddles two distinct geological regions: the flats and filled marshlands of lower Roxbury and North Dorchester and the highlands of outer Roxbury, with steep hills and serpentine dales. Roxbury gets its name from the rock outcroppings, known as puddingstone, that dot the landscape. A conglomerate form of rock, puddingstone is made up of large pebbles, cobbles, and rounded or angular rocks and boulders that appear to float in a matrix of fine-grained material. Known by geologists as Squantum tillite, the puddingstone is as old as 600 million years.[1]

Roxbury began in the 1700s as a small farming village, like so many other early New England settlements, stretching out along the Boston neck, the only all-land route to the colonial city of Boston, a spindly, hilly peninsula. The Dudley area, which gets its name from the street that runs from Dudley Station to Uphams Corner (Dudley Street is named after Thomas

Dudley, a seventeenth-century governor of the Massachusetts Bay Colony), abuts the South Bay, once a glistening tideland filled earlier in the twentieth century to accommodate the city's growing population and industrial sector. On a wet, windy day, a person in Roxbury can still smell the sea in the gusts that blow from the northeast, a reminder of the area's maritime past. The Shirley-Eustis House, the colonial governor's mansion and now a National Historic Landmark, sits high atop a knoll in lower Roxbury, just off Dudley Street, taking full advantage of the eastern views toward the South Bay and downtown Boston just beyond.

Before 1850, Boston's geography stifled large-scale expansion. The system of rivers, marshes, and bays restricted pedestrian communication. The deepwater harbor, on the eastern and northern part of the Boston peninsula, confined the city. For this reason, land had always been expensive, and as the city's economy and population grew, developers and engineers dammed and filled many of Boston's waterways and wetlands, first for commercial, and later for residential purposes. By 1900, the South End and Back Bay, once wet, were completely settled.[2] The South Bay, too, was filled around this time.

The combined forces of industrialization and immigration fueled nineteenth-century Boston's economy, transforming its landscape and population. Prior to the 1840s, Boston was comprised mainly of farmers, artisans, and millworkers from the British Isles, as well as immigrants from the Canadian Maritime Provinces and rural New England. After the 1840s, waves of Irish, Jewish, and Italian immigrants arrived in Boston, settling throughout the city in well-delineated ethnic precincts. Roxbury became home to substantial numbers of Irish, as well as Jews. In addition, according to 1865 census data, eighty-three blacks lived in Roxbury and Dorchester.

An independent town until 1868, Roxbury, along with the nearby towns of Dorchester (annexed to Boston in 1869) and West Roxbury, were Boston's first suburbs, attracting middle-class residents eager to escape the noise and nuisances associated with an industrialized city. By the middle decades of the nineteenth century, the farms of lower Roxbury gave way to manufacturing and commercial uses, and also to housing for workers. Access to nearby water resources such as the Stony Brook and Jamaica Pond, and to the wharves of the South Bay, and its proximity to downtown

Boston encouraged the industrial development of the Dudley area. Meanwhile, the Roxbury highlands remained primarily residential in character, owing to the area's hilly terrain.[3]

By the turn of the century, lower Roxbury experienced a shift from a lower middle-class to a working-class district, as many families who could afford to move abandoned the area for the new suburbs farther to the west and south. Similarly, industrial changes such as consolidation and increased scale of business left Roxbury with scores of vacant buildings as old factories closed or moved to areas with more space to expand. According to historian Sam Bass Warner, by the early 1900s, "lower Roxbury was becoming a slum."[4]

The availability of cheap, undeveloped land and improvements in transportation infrastructure such as the expanded street railway system enabled the middle class to flee Boston's deteriorating old neighborhoods late in the nineteenth century. Inspired by the rural ideal of private family life, a small community setting, and access to natural areas, modern suburbs emerged by the early 1900s.

Notwithstanding the middle-class suburban flight, the Dudley area maintained a working-class profile well into the twentieth century, with Irish, Italian, and Jewish residents. Beginning in the 1940s, Roxbury experienced a significant increase in its black population, due to the migration of southern blacks who were being displaced by the mechanization of agriculture. The black population grew from 5 percent in the Dudley area in 1950, to 20 percent in 1960 and 50 percent in 1990. Among the black immigrants were Cape Verdeans, who arrived from their West African homeland in large numbers in the 1960s. During the 1960s and 1970s, the Dudley area's Latino population—Puerto Ricans, Dominicans, Hondurans, Guatemalans, Cubans, and Mexicans—expanded, growing from 12 percent in 1970 to 28 percent in 1980 and 30 percent in 1990.[5]

As people of color began settling in Roxbury, whites moved out in droves. Between 1950 and 1980, the Dudley area's population declined by more than half. The white population dropped from 95 percent in 1950, to 79 percent in 1960, 45 percent in 1970, and 7 percent in 1990.[6] In just two years, 1968 to 1970, tens of thousands of Jews left Roxbury and Dorchester, ending fifty years of settlement in the area. The once-bustling business district along Dudley Street became a wasteland by the late 1960s as the number of private businesses dropped from 129 in 1950 to 49 in 1970, most of them small bodegas, restaurants, and auto body shops. By 1988, banks had closed 40 percent of their branch offices in Roxbury. Manufacturing jobs declined from 20,000 in 1947 to 4,000 in 1981, when the unemployment rate in Roxbury reached 12 percent, twice Boston's overall rate. Meanwhile, communities along Route 128, a beltway ten miles outside Boston, boomed, outpacing Boston's per capita income every year since 1959.[7]

The white flight the Dudley area experienced occurred throughout urban America in the three decades between 1950 and 1980. All across America, as the suburbs flourished, the cities languished. Private and public investment in schools, housing, businesses, and physical infrastructure came to a halt, the result of the wholesale relocation of the middle class from urban areas to the suburbs. Discriminatory practices in the real estate, banking, and insurance industries, so-called red-lining, left residents in inner-city neighborhoods without credit or insurance, further segregating them from the rest of society. Federal subsidies for mortgages and highway construction promoted suburban growth over urban revitalization. Cheap, raw land lured droves of businesses and developers from the central cities to new suburban office parks and subdivisions.

The urban renewal programs of the 1960s and 1970s were an ill-conceived response to white flight. In Boston, working-class, multicultural neighborhoods, such as the South End, were redeveloped, displacing tens of thousands of lower-income individuals unable to afford the price of gentrification. Thousands of acres of dilapidated housing were demolished and rebuilt, attracting a new professional class to formerly working-class districts.

In the Dudley area, urban renewal, white flight, red-lining, and disinvestment resulted in a literal holocaust. Despite the need for affordable housing, building owners, developers, and speculators set fire to hundreds of properties in Roxbury in the 1970s and 1980s to drive out lower-income residents, gut the buildings, and collect insurance proceeds to pay for rehabilitation. The Highland Park neighborhood, which abuts the Dudley area, was deemed the arson capital of the nation in 1981.[8] Unable to secure mortgages or financing for home improvements, many in the Dudley area abandoned their houses, as the number of absentee owners and vacant lots grew exponentially. Between 1947 and 1976, 648 buildings—nearly half the area's housing stock—were demolished in Roxbury. In addition, as a result of the conflagration and demolition, approximately 840 vacant lots covering 177 acres were scattered about the Dudley area. These lots attracted all sorts of unwanted, if not illicit, activities, from midnight dumping of hazardous waste to drug dealing, and they served as a breeding ground for rats. The Boston Redevelopment Authority, evaluating neighborhood conditions in the 1970s, warned that "abandonment, if allowed to proceed unchecked, can spread like cancer, taking whole city blocks."[9]

Dudley area resident Paul Bothwell, a white Baptist minister who moved to Roxbury from the South End in 1976, witnessed firsthand the deterioration and blight that afflicted the neighborhood. "It's not something that's just sheer chance. It's not because people are stupid," Bothwell states. "This was really part of much, much larger forces that are at work and they may or may not be consciously malicious. . . . This is the result of . . . large-scale things that systematically cripple or dismember a community." Recalling his and his neighbors' desperation and drive to improve the Dudley area, Bothwell points to the vacant, blighted land as the focus of early revitalization efforts. He remembers, "[The City of Boston was] always tearing things down. . . . We were trying to build some sort of a land trust . . . that would place control of the land in the hands of community people."[10]

As it happens, the city put Bothwell and his neighbors in touch with an official from the U.S. Department of Agriculture, mistakenly believing that the Dudley residents simply wanted to develop a community garden. Instead, the official helped the residents engage in a comprehensive planning process for rebuilding their devastated neighborhood through reclaiming the blighted land as a potential asset for developing housing and parks, in addition to gardens.

After several years of stalled revitalization efforts on the part of community residents and local agencies, including leading community organizations like La Allianza Hispana and Nuestra Comunidad, and a handful of planning reports prepared by consultants that outlined the Dudley area's many problems, the Mabel Louise Riley Foundation, one of Massachusetts' leading private foundations, intervened in 1984. The foundation's trustees saw the neighborhood as the most disadvantaged area in the city; the area's blighted landscape, symptomatic of larger social and economic decline, affected the trustees in an immediate and visceral way. Consequently, they engaged local stakeholders in the formation of the Dudley Advisory Group, charged with the mission of establishing a new organization to rebuild the area's damaged social, economic, and physical fabric.

In 1985, the Dudley Street Neighborhood Initiative (DSNI) was born, with the mission "to empower Dudley residents to organize, plan for, create and control a vibrant, high quality and diverse neighborhood in collaboration with community partners." Almost immediately, this bold purpose was reduced to specific, land-related activities. In early 1986, responding to resident priorities, DSNI created an organizing campaign, Don't Dump on Us, led by DSNI organizer Ros Everdell. The goal of the campaign was to clean up more than thirteen hundred blighted lots, over fifty of them confirmed hazardous waste sites, in the Dudley area and, in the process, to unify residents, businesses, and agencies in the common pursuit of neighborhood improvement. Over a hundred residents participated in a neighborhood cleanup in the summer of 1986, and the city set up a new hot line, 725-DUMP, to report illegal dumping. In 1987, DSNI targeted two illegal trash transfer stations, which were notorious for attracting vermin and causing putrid odors, especially on hot summer days. Working with the city's public health department, residents shut down the facilities. According to DSNI member Father Walter Waldron of St. Patrick's Church, closing down the trash transfer stations emboldened DSNI: "I think when some of the trash folks were trashed—closed down—that it then became [acceptable] to believe that if they could be changed, well, then [so could] other things."[11]

The Don't Dump on Us campaign was the start of DSNI's efforts to promote environmental justice in Roxbury and served as the basis of its community-building strategy. The campaign symbolized the critical link between the Dudley neighborhood's struggle for environmental quality and its effort to promote social and economic justice. DSNI conceived the community and environment as one and the same. The visibility and scale of the neighborhood's physical degradation made environmental action the most powerful lever for addressing overall community revitalization.

Uplifted by the euphoria surrounding their success in dealing with the trash transfer stations, DSNI instituted a community planning process in 1987 with the help of planning consultants from DAC International, a Washington,

D.C.–based firm. This planning process resulted in "The Dudley Street Neighborhood Initiative Revitalization Plan: A Comprehensive Community Controlled Strategy," a master document setting forth the principles that would guide the Dudley area's development. At the core of the document is the notion of an urban village: a diverse, economically viable, and neighborly community that would combine housing, shopping, parks, and a community center, each aimed at fostering a sense of place and safety.

The following year, DSNI became the only community-based nonprofit organization in the country to be granted eminent domain power over abandoned land within its borders, a power DSNI recognized as critical to its success in reclaiming the neighborhood. Eminent domain gave DSNI the ability to gain ownership and control over key parcels of privately owned land for development in accordance with DSNI's plans for an urban village. With a planning document and eminent domain power in hand, DSNI was ready by the end of the 1980s to begin rebuilding the neighborhood, literally from the ground up.

Back to the Future: Agriculture as the Key to DSNI's Urban Village

The land, so important to bringing about the formation of DSNI, became the primary vehicle by which DSNI served its mission. DSNI realized that vacant lots were a liability that could be converted into a powerful asset, like affordable housing for area residents. The land became an emblem of the neighborhood's troubled past and its new direction. In 1994, DSNI's first housing development, Winthrop Estates, a shining example of this new direction, was completed to great fanfare, a milestone in the Dudley area's revitalization. But beyond housing, how would DSNI transform the neighborhood's countless vacant lots into a thriving urban village? How would the land be converted from a liability to an asset comprehensively and on a large scale? The answer would come from Roxbury's past: its agricultural heritage.

By the mid-1990s, the Dudley area still contained an extraordinary amount of vacant land, much of it contaminated. According to the Massachusetts Department of Environmental Protection, the Dudley neighborhood contains at least fifty-four hazardous waste sites, with dozens, if not hundreds, more suspected to be contaminated. Common pollutants include lead, chromium, mercury, asbestos, and especially petroleum constituents like polyaromatic hydrocarbons and benzene. These hazardous substances are the typical by-products of the industrial facilities that for decades called Roxbury home, such as metal fabricators and electronics manufacturers. Oil-based pollutants come from leaky underground storage tanks, auto body shops, and gas stations, typical sources of oil spills in urban areas. Pollutants associated with pesticides like arsenic are also widespread, toxic evidence of the apple orchards that were part of the neighborhood's agricultural past.

The environmental problems afflicting the Dudley neighborhood have fallen through the cracks of traditional environmental laws. The state's hazardous waste site cleanup law, Chapter 21E of the Massachusetts General Laws (21E), was modeled largely after the federal Comprehensive Environmental Response, Compensation and Liability Act—the Superfund. Like the Superfund, 21E requires the parties responsible for contaminating land and groundwater to pay for the full cost of cleanup, even if a party only minimally contributed to that contamination. As well, banks and other secured lenders that provide financial services to the owners or operators of contaminated sites can potentially be held liable for cleanup. Further, even after a cleanup is undertaken, parties can still be held responsible to third parties, such as neighbors or abutters, for personal injury or property damage. In effect, these laws provide no end point to liability, adding uncertainty to an already costly process.

Although since originally enacted in the early 1980s both 21E and Superfund have been revised to make them less burdensome on potentially responsible parties, they are nonetheless among the blunter instruments of environmental law, intended to compel quick response on the part of polluters while deterring future spills or releases of hazardous materials such as oil, lead, asbestos, and other substances commonly used in commercial and industrial operations. Rather than clean up contaminated sites, parties have typically chosen to litigate the issue of liability, dragging in other potentially responsible parties. At the same time, they have abandoned contaminated sites, like those found in Roxbury and many urban areas, in favor of clean, cheaper land in suburbs and rural areas.

These laws essentially provide disincentives for parties, whether causally responsible for polluting a site or not, to report and remediate contamination. This is why, of the hundreds of thousands of confirmed contaminated sites nationwide and the roughly eleven hundred sites in Massachusetts, over fifty of them in the Dudley neighborhood alone, only a small fraction have been cleaned up. Most languish untouched.

Thus, an unintended consequence of the hazardous waste site cleanup laws in Massachusetts, as in the rest of the country, has been the development of raw suburban and rural land, accelerating the rate of sprawl and destruction of habitat and open space while further abandoning the inner city. In addition, these laws have done little to encourage businesses to seek out nonhazardous materials for use in their day-to-day operations, ensuring the prevention of future contamination. Most firms would simply rather comply with existing environmental regulations, and avoid hazardous waste sites, than adopt pollution-prevention strategies.

Other land use–related problems in Roxbury have gone unaddressed by traditional environmental laws. Trash transfer stations and other unwanted land uses, such as auto body shops, that contribute to environmental degradation in Roxbury usually fall under the radar screen of most environmental laws. Because they are typically small facilities, these businesses do not require any comprehensive environmental review under the Massachusetts Environmental Policy Act, the state statute requiring the assessment of environmental impacts for projects that trigger numerical review thresholds, based on somewhat arbitrary estimates of the causal connection between a projected quantity of pollution and its effect on specific media or natural resources. Instead, they are treated as run-of-the-mill land use decisions made by the city's zoning board, which lacks any formal environmental review authority. When environmental problems at these facilities do arise—whether odors, noise, or noxious air emissions—local public health officials from the City of Boston Office of Environmental Health, not the state Department of Environmental Protection, are in the best position to respond. Acting on their power to protect the public health dating back to the nineteenth century, these officials are in fact the real environmental regulators when it comes to neighborhoods like Dudley.

Environmental laws have thus historically been of little value to Dudley residents. In the case of contaminated sites, 21E has failed to encourage cleanups in the neighborhood. Instead, it has helped push development opportunities outside the city. At the same time, unwanted land uses like trash transfer stations have been allowed to pile up in Roxbury because they have been dealt with on a case-by-case basis, without any comprehensive environmental review, and because such uses tend to follow the path of least resistance.

Nearly 21 percent of the land in the Dudley area today is blighted and abandoned, a stagnant reminder of the conflagrations and white flight of earlier decades. The tell-tale topography of America's inner-city neighborhoods, these sites, many of them contaminated brownfields, continue to dominate the landscape. Even in the strong economic climate of the late 1990s, the Dudley area remains one of the hardest-hit neighborhoods in Boston. The diverse community of African American, Latino, Cape Verdean and white residents has a per capita income of just $7,600, compared to nearly $16,000 for the city as a whole. The median family income for the area is $20,848, and the unemployment rate is roughly 16 percent. Approximately 32 percent of the Dudley neighborhood's population falls below the poverty level.[12]

On top of these neighborhood deficits, lead poisoning is a major health threat in the Dudley area, as it is in many other lower-income urban communities. Nearly every street has at least one confirmed case of childhood lead poisoning from old paint or contaminated soil and water. In response to this pervasive environmental crisis, DSNI launched a lead poisoning prevention project in 1994, under the stewardship of Ros Everdell and Trish Settles, the newly hired environmental organizer. Although DSNI had been working on environmental issues since its inception, culminating in the establishment in 1991 of Dudley PRIDE (People and Resources Investing in Dudley's Environment), a grass-roots campaign to promote environmental health and government responsiveness to neighborhood environmental problems, Settles became DSNI's first in-house environmental professional, with experience as a hydrogeologist in northern California and two advanced degrees in environmental science and policy. Part organizer, part technical adviser, Settles took the lead in working with residents to tackle the lead poisoning problem.

At the same time, DSNI started to work with area gardeners on lead abatement strategies as a way of reducing the risk of lead exposure. For decades, residents had been cultivating backyard gardens with little or no awareness of the hazards posed by soil contamination from lead and other pollutants. Settles and Everdell realized that local gardeners were a natural constituency for disseminating information about lead poisoning and ways to prevent it. They set about forming a network of gardeners. In 1995, with a planning grant from the Nathan Cummings Foundation, they explored the possibility of converting this network into a base for local food production, or urban agriculture, not only to combat environmental contamination like lead but to bolster the local economy. If residents were able to sell their crops, they could possibly earn a living from it.

Urban agriculture, or "urbaculture," is an ancient concept. For centuries, food sold in cities around the world has been grown locally. In Chinese cities, 90 percent of the vegetables is locally grown. In Africa, 20 percent of total food production occurs in cities. Urban farmland consists of diverse sources: in Calcutta, the garbage dumps; in Mexico City, the rooftops; in Jakarta and Nairobi, dirt strips along the road.[13] In the United States, many cities have set aside land for gardens to be used by local residents. One of the oldest is found in Boston, in the Fenway section, part of Frederick Law Olmsted's Emerald Necklace park system. Since Olmsted's day, gardens have proliferated in American cities, especially over the past several decades as part of the movement to "greenify" cities.

Seeking to build on this urban agriculture tradition, a new nonprofit organization, the Food Project, began recruiting Roxbury youth in 1995 to help harvest crops from their land at Drumlin Farm, an environmental education center run by the Massachusetts Audubon Society and located in the Boston suburb of Lincoln, Massachusetts. The Food Project operated a farmers' market in Roxbury, selling fresh produce to area residents during the summer. The Food Project's directors, Pat Gray and Greg Gale, reached out to Settles in hopes of collaborating with DSNI on their project. The timing was perfect.

What is more, DSNI had been the beneficiary of a major enforcement action brought by the New England regional office of the EPA against the Massachusetts Highway Department (MHD) for illegally disposing of hazardous waste at several MHD equipment yards across the state. As part of its settlement with EPA, MHD agreed to fund a Supplemental Environmental Project (SEP) for select sites around Massachusetts. An SEP is an alternative form of penalty for environmental violations; rather than simply levying a fine, which is paid to the federal government, it allows the violator to fund local initiatives that are related to the issues underlying its violations. Officials from EPA/New England approached DSNI to see if there might be an appropriate SEP site in the Dudley area, preferably a contaminated site that DSNI wanted to clean up for environmentally related purposes. Without hesitation, Settles pointed out a site on the corner of Dudley Street and Brook Avenue, a former garage and auto body repair business, located not twenty yards from DSNI's Dudley Street offices. The Brook Avenue site had been identified by residents as an important brownfield and a prime location for a possible greenhouse or bioshelter, which could support area gardeners and provide a year-round growing season.

Greenhouses are structures that use large glass or plastic panels to collect and trap solar energy for growing fruits, vegetables, herbs, and flowers. Bioshelters are similar to greenhouses in form; however, they include additional uses, such as systems to cultivate fish stocks and water-borne plants, known as aquaculture. The water used to raise aquatic plants and animals provides an extra thermal mass that increases the bioshelter's energy efficiency. Their design makes year-round food production economically viable and reduces the need to use fossil fuels (which contribute to global climate change) for heating. EPA and MHD approved the greenhouse-bioshelter SEP in 1995, fueling the momentum toward an even more ambitious brownfields and urban agriculture project.

Perhaps what really clinched the case for urban agriculture was the hiring in 1995 of Greg Watson as DSNI's new executive director. Watson replaced Rogelio Whittington, a local resident, respected community leader, and accountant who oversaw the successful Winthrop Estates project and DSNI's emergence in the early 1990s as a national model for resident-driven community planning and revitalization. Like Settles, Watson brought to DSNI an abiding interest and expertise in environmental issues, cutting his teeth as a founding member of the New Alchemy Institute, an environmental education organization, and later as commissioner of the Massachusetts Department of Agriculture under Governor Michael Dukakis.

Watson's stewardship, coupled with the lead poisoning project, a gardeners' network, Food Project collaboration, and the greenhouse-bioshelter SEP, led to a powerful synergy around brownfields cleanup and urban agriculture, which influenced a series of community visioning sessions held throughout 1996-1997 involving more than two hundred residents. These sessions were a continuation of the DAC planning process initiated in 1987. In the sessions, residents, businesspeople, and representatives from local nonprofit organizations engaged in a dynamic process of visioning and planning around the concept of the urban village, identifying community assets and resources, as well as problems and deficits. They looked at every facet of the community, from the economy and crime, to arts and culture, to education, and the environment, as part of building the urban village. The process also gazed into the future to anticipate the changes, both inside and outside the community, that would likely affect the Dudley neighborhood. For example, in discussing the feasibility of urban agriculture, the participants explored the likely impact of global climate change on the local climate and growing conditions.

Urban agriculture emerged from the visioning sessions as a key strategy in the effort to restore not only a healthy, vibrant local economy but a clean, high-quality environment, each central to DSNI's urban village. In the spring of 1997, the Urban Agricultural Strategy (UAS) was launched, with major funding from the Ford Foundation. The UAS's mission is to clean up the dozens of brownfield sites and vacant lots in the neighborhood and redevelop them for food production and value-added food enterprises such as prepared foods—smoked fish, preserves, dried mushrooms.

According to Watson, the UAS represents a comprehensive approach to community building, linking overall community health to economic development:

> Residents will be empowered by realizing how much control they can have over meeting one of their very basic needs. Local farms, gardens and greenhouses will benefit children and families by providing them with more access to fresh, nutritious, locally grown foods—so essential to proper development of infants; urban agriculture will make a significant contribution to our local, village-scale economy and, by focusing on abandoned lots and brownfields, gardens, farms and greenhouses will play a key role in transforming current liabilities into community assets.[14]

The UAS is designed to put an end to Roxbury's and Boston's dependence on mammoth agribusinesses for their food supply. On average, a single American farmer today grows enough food each year to feed one hundred people. However, as the author Michael Pollan warns, such productivity has come at a substantial price: "The modern industrial farmer cannot achieve such yields without enormous amounts of chemical fertilizer, pesticide, machinery, and fuel, a set of capital-intensive inputs . . . that saddle the farmer with debt, threaten his health, erode his soil and destroy its fertility, pollute the ground water and compromise the safety of the food we eat."[15] Watson wants to replace industrial or corporate agriculture with living analogues of natural systems—community gardens, greenhouses, and bioshelters—powered by renewable energy sources such as solar, wind, and hydrogen. In the tradition of organic agriculture, the UAS seeks to substitute a diverse, pesticide-free agricultural system based on nature's model for the large-scale, pesticide-driven monoculture of agribusiness.

He notes that more than 85 percent of the food consumed in Massachusetts is imported; nearly 60 percent of that is grown in California. "We are therefore dependent on a source 3,000 miles away to supply us with a significant portion of our food," he states. The implications are serious: Massachusetts residents pay between 6 and 10 percent above the national average for their food. Moreover, food shipped over long distances loses much of its nutritional value, not to mention the environmental costs associated with transport. "It is become increasingly evident that California will not be able to continue growing and exporting food at the current rate much longer," Watson warns. "The highly mechanized, resource-intensive farming methods are undermining the ecological infrastructure that supports the agricultural economy."[16]

For Settles, the UAS's power stems from its environmental core. "Environmental cleanup is an issue that everyone can get their hands around," she explains, "but we've had to reclaim the notion of what the environment is, just as we've had to reclaim the environment itself from pollution and abuse." Settles is quick to point out that the UAS's combination of a food- and environment-based strategy "connects people in a fundamental way."[17] It is accessible to Dudley residents, especially those immigrant groups like the Cape Verdeans who come from rich agricultural backgrounds.

In explaining what the UAS will look like, Watson refers to a description of urban agriculture put forward by John Todd and George Tukel, two urban planning and design pioneers. They suggest that urban agriculture will have many forms:

> Shade trees will be partially replaced by an urban orchardry of fruits and nuts. Sunlit walls will become architectural backdrops for espaliered fruits and crops. Shrubs,

which purify air by removing auto exhaust, lead and zinc will be planted in raised beds between the streets and sidewalks. Community gardens and gardening will increase as participation grows. Agricultural bioshelters will fill vacant lots and ring parks. Floating bioshelters will line harbors and produce their fish, vegetables, flowers and herbs for sale. Old warehouses and unused factories will be converted into ecologically inspired agricultural enterprises. Fish, poultry, mushrooms, greens, vegetables, and flowers will be grown in linked and integrated cycles. Roof tops will utilize bioshelter concepts for market gardens all year.[18]

The UAS envisions a food system that includes a community-supported farm, a network of community gardens, several greenhouses and bioshelters, processing facilities, and a variety of marketing outlets, such as restaurants, grocery stores, farmers' markets, and a web site on the Internet. With the community controlling most of the means of production, distribution and marketing will be handled by a mix of private and community businesses.

The plan for the Brook Avenue greenhouse-bioshelter, for example, contemplates growing high-value crops such as lettuce, tomatoes, and cucumbers without the use of chemical pesticides, herbicides or fertilizers. The greenhouse-bioshelter will employ organic production techniques, as will all UAS businesses. The crops will then be sold to high-end markets in the Boston area, such as restaurants and corporate cafeterias.

Trees and urban orchardry are another component of the UAS. Watson views trees as among the most underused resources in urban communities. "The potential trees have for providing food, moderating temperature, improving air quality and scenic vistas cannot be overestimated," he declares. Several Dudley residents have suggested reintroducing the Roxbury russet apple, a famous crop of earlier decades, to the area and processing it into value-added products like cider and apple butter. The Bartlett pear, another possible crop candidate, was developed in Roxbury in the nineteenth century.

In the light of the agricultural heritage of many of Dudley's residents, the UAS provides a unique opportunity for cultural expression. As Settles sees it, residents will be able to affirm their personal and cultural identity, thus contributing to the urban village's unique and powerful sense of place and creating new market niches for products such as agriao, arruda, yerba, and yellow yams. Sustainable, organic agriculture is not a novel idea to many Dudley residents, who come from farming traditions where expensive chemical inputs and machinery are limited.

A critical part of the UAS is the development of markets for Roxbury's agricultural products. With the food industry dominated by global corporations whose primary interest is the bottom line, variety, freshness, nutritional value, and affordability are hard to come by. DSNI wants to give local residents and businesses greater control over the food marketing system.

One approach is a farmers' market. In Boston, farmers selling their products directly at local markets are able to earn between 10 and 25 percent more for their produce than at commercial markets, while consumers can save between 5 and 10 percent. This is possible because farmers' markets eliminate the need for middlemen, such as packagers, truckers, wholesalers, distributors, and supermarkets.

Community-supported agriculture (CSA) is another approach for direct marketing. In the CSA model, consumers become partners in the farm from which they purchase produce. Consumers prepurchase a portion of their produce during the winter when farmers are planning for the coming spring and when cash flow on the farm is usually low. Value-added products such as preserves, sauces, and baked goods also provide an opportunity for the UAS to compete in the agricultural marketplace. They add to the community's marketing outlets and enable Dudley farmers and gardeners to retain a larger portion of the money spent on their products.

In developing a business strategy, Watson and his DSNI colleagues have looked to Harvard Business School professor Michael Porter's study, "The Competitive Advantage of the Inner City."[19] Porter finds that some inner-city communities enjoy a natural competitive advantage with respect to certain industries, including food-related businesses. He concludes that strategies for developing specialty retail niches remain largely unexplored by many urban communities and offer a potentially significant business opportunity. The UAS is thus designed to capitalize on Roxbury's natural competitive advantage by bringing to the market products that cater to the unique tastes and preferences of Boston's diverse population.

To achieve economic viability, the UAS will require intensive crop production systems, including intercropping and successional planting. Such strategies place tremendous demands on soil fertility, which requires ongoing soil-building activities like composting and remineralization (the restoration of inorganic micronutrients). Yet given the importance of the soil to the UAS, DSNI must inevitably reckon with the neighborhood's countless polluted sites

The Brook Avenue greenhouse-bioshelter is the UAS's pilot brownfield site. The SEP subsidized the cleanup and redevelopment of the site, but it also provided DSNI with an opportunity to learn how to go about transforming blighted, vacant land into productive agricultural uses. The feedstock and foundation of the UAS, brownfield sites like the Brook Avenue lot present a host of challenges that must be anticipated and resolved if the UAS is going to achieve the large-scale success the urban village requires. As Greg Watson notes, the UAS is all about "transforming brownfields into greenfields and greenhouses." However, he explains, "Some of the brownfield sites may not be suitable for growing crops after remediation. They may . . . still be incorporated into our urban agricultural system. Some could become the future site for a community food processing facility, aquaculture farm, mushroom factory, greenhouse or other kinds of food-related businesses."[20]

In addressing the challenge posed by brownfield redevelopment, DSNI must first identify appropriate brownfield sites like the one at Brook Avenue that can be marketed to developers and users according to the specifications and requirements of the UAS. These sites must be evaluated not only in terms of environmental conditions but ownership. A great challenge for the UAS and any other large-scale brownfield effort is to gain access to or otherwise generate information about brownfield sites. DSNI must first determine who owns the sites. If a site is publicly owned, by the City of Boston, for instance, then DSNI can negotiate with the city's Department of Neighborhood Development (DND) to gain control of the site or to dispose of it to a third party. This must occur in compliance with the applicable public bidding regulations that govern the sale of all public property. If a site is owned by a private owner, then DSNI must identify the owner and the tax status of the property. If the owner has paid all property taxes and there are no other liens on the property, DSNI must attempt to convince the owner to sell the property or develop it as part of the UAS.

Typically private owners of sites in neighborhoods like Dudley are hard to contact, let alone negotiate with. They often live outside the city, have unlisted telephone numbers, or operate under a variety of business names. Usually they are speculators, waiting for the right opportunity to sell their land and any buildings on it for maximum profit. If the owner is delinquent in payment of property taxes, the city can exercise its power to take the property through foreclosure proceedings. Because of concerns about liability for nuisance and environmental hazards, municipalities are usually averse to foreclosing on properties in inner-city neighborhoods in the absence of a compelling redevelopment plan or strong political pressure.

Another hurdle that must be surmounted with private property is access to the site to conduct environmental assessment. Without the express permission of the owner or an imminent threat to human health and the environment caused by on-site conditions, neither DSNI nor any other other entity can legally gain access to the site. Such illegal access would constitute a trespass. The upshot is that privately owned sites are difficult to incorporate in a brownfield redevelopment strategy like the UAS.

Even assuming that the issue of site ownership can be effectively managed, the matter of environmental assessment remains. Whether the land is publicly or privately owned, most owners are reluctant to proceed with site assessment work in the light of their fear of potential legal liability should they uncover reportable or significant contamination levels. The preferred approach is blissful ignorance. Yet municipal officials can be persuaded to conduct site assessments and cleanups, especially if they understand that the risks of going ahead with an assessment and cleanup are not much greater than the risks associated with simply owning a contaminated site. Further, the hazardous waste site cleanup laws in Massachusetts, as in most other states, already provide certain liability protections for municipal owners who did not cause or contribute to contamination and who attempt to dispose of a brownfield site. The new Massachusetts brownfields bill, enacted into law in the summer of 1998, provides additional exemptions from liability so as to encourage municipalities to conduct site assessments and cleanups and to market sites.

Despite incentives for brownfield redevelopment, hard-hit urban neighborhoods face challenges beyond the complexities of environmental cleanup. Market forces often overcome even the most daunting of environmental problems, as long as the location of a brownfield site is deemed desirable. *Desirability* itself is really a code word—a proxy for the often masked concerns of real estate brokers, developers, and municipal officials about the relative safety, economic viability, and racial composition of neighborhoods that host brownfield sites. Social segregation, racial discrimination, and misperceptions about crime, poverty, and the existence of a skilled labor market in inner-city communities are often determinative factors in development decision making. Brownfield sites in these communities are usually ruled out well before any serious consideration is given to environmental issues. Brownfield redevelopment is thus at least as much a function of the perceived social and economic conditions in a community as it is the perceived contamination on a single site.

This means that for brownfields redevelopment to succeed as part of a larger community revitalization strategy, neighborhoods like Roxbury deemed too risky by the market must aggressively promote themselves and defy media-driven stereotypes and the ravages of a legacy of disinvestment, white flight, and racism. They must demonstrate that many "less desirable" locations are in fact chock full of law-abiding, educated consumers, workers, and managers, just like any other neighborhood.

Still in its early stages, the UAS must reckon with the knotty problems of landownership, contamination, and marketability. The best approach appears to be to identify publicly owned sites, like the Brook Avenue site, and, working closely with the DND and the Boston Redevelopment Authority, to establish an assessment and marketing plan in accordance with the UAS's mission. With the power of eminent domain at its disposal, DSNI itself might consider taking certain properties, but it too must consider the potential liability and the responsibility for maintenance associated with ownership of any neighborhood site.

The Brook Avenue greenhouse-bioshelter, which will be finished by the end of 1999, is a beacon for future UAS efforts and has jump-started the food production and revitalization process. Planning is currently underway to determine exactly what kind of crops the greenhouse-bioshelter will grow. DSNI hopes that within a decade, the UAS will transform the blighted, economically distressed Dudley neighborhood into a thriving urban agricultural center, where residents and local businesses contribute directly to the local economy and serve as stewards of the environment. DSNI's vision is to achieve recognition as a center of agricultural production on a par with places like Vermont or Maine. Stickers will identify produce such as greens, tomatoes, corn, and herbs, and value-added products like jams, dried mushrooms, and smoked trout as "Made in Roxbury." This brand identification, it is hoped, will enhance the marketability and competitive advantage of Roxbury-grown products.

Civic Environmentalism and Urban Agriculture

DSNI's experiment in urban agriculture, though still in its early stages, provides a powerful example of civic environmentalism in action. Out of extraordinary adversity spanning several decades, the Dudley neighborhood has been able to pull together to engage in comprehensive community planning, resulting in a sophisticated, innovative community-building strategy. With the full participation of neighborhood residents, businesses, community organizations, and local foundations, DSNI arrived at the UAS slowly and deliberately, improving its civic capacity for planning and problem solving along the way, creating what Greg Watson calls "civic alchemy"—the yeasty, creative mix that comes from an engaged citizenry and produces often unpredictable, though always beneficial, results. DSNI recognized the primacy of environmental quality to the neighborhood's overall health and was able to integrate environmental cleanup and conservation-based strategies into their comprehensive revitalization plan.

DSNI's planning process became the vehicle for educating local stakeholders about environmental issues and solutions, culminating in the development of the UAS. Moreover, with environmental justice as a constant theme, DSNI was able to merge traditional civil rights and social justice concerns, so dear to most Dudley residents, with a call for environmental protection. This helped to draw in local stakeholders who might otherwise not be interested in a purely environmental project. Having lived with environmental harms for decades, the Dudley community was eager to embrace the UAS's environmental mission.

While in many respects dependent on and constrained by the laws governing hazardous waste site cleanup, land use, and property rights, the UAS is attempting to move beyond the bounds of conventional law by creating an overarching community revitalization plan that, it is hoped, will allow DSNI to initiate brownfield redevelopment on a broad scale. In essence, DSNI's plan, not the law, is driving the UAS, though state and local regulations will inevitably affect it.

The UAS is a compelling example of industrial ecology. Grounded in the principles of sustainability and pollution prevention, the UAS incorporates strategies designed to convert contaminated land into safe, productive agricultural uses that incorporate renewable energy and organic production methods. As part of the urban village model, it looks to create a local economy that actually adds to, rather than detracts from, the community's physical conditions. Gardens, orchards, greenhouses and bioshelters, and street trees will give the Dudley neighborhood the appearance of a thriving natural area of the kind typically found only in more rural settings. At the same time, these ecological assets will be the drivers of the area's economy, providing jobs and creating wealth for local residents. The UAS is a prime example of industrial ecology as sustainable land use.

Greg Watson, Trish Settles, and others in the neighborhood have also discussed the possibility of developing an ecoindustrial park in the nearby South Bay industrial area. The South Bay is adjacent to the Dudley neighborhood and has long been home to a variety of polluting facilities, like the South Bay Incinerator, abandoned for over twenty years, and numerous junk yards and auto body shops. DSNI is among a number of community groups from Roxbury, Dorchester, South Boston, and the South End calling themselves Neighborhoods United for the South Bay (NUSB) that in 1995 convinced the state's Division of Capital Planning and Operations, the owner of the incinerator site, to demolish the facility and work with NUSB to devise an environmentally sound redevelopment plan. DSNI would like to see an ecoindustrial project that complements the UAS, and with NUSB has taken up the cause. Potential end uses include industrial recycling and remanufacturing businesses. The ecoindustrial park proposal is just beginning to gain momentum, as the fate of the site hangs in the balance.

The UAS will also contribute to a renewed sense of place by greening and improving the landscape while giving local residents a stake in the quality of their physical surroundings. Instead of a source of blight and decay, the land will become a symbol of pride, harmony, and community. The UAS will soften the edges of what is now a hard and forbidding inner-city environment and allow residents and visitors the opportunity to experience Roxbury as a living ecosystem and vibrant community.

The UAS will have to surmount formidable obstacles if it is to realize its promise as a model of civic environmentalism. First, DSNI must identify and recruit residents and businesses prepared to engage in organic, state-of-the-art food production. This will require training and consultation with experts in the field of urban agriculture. In addition, DSNI must help UAS participants overcome the legal and financial challenges posed by brownfield redevelopment, such as ensuring protection from liability for contamination caused by former users and obtaining low-cost loans and grants for assessment and cleanup activities. Assistance from local lenders and from the funding mechanisms set up by the state's brownfield legislation is key to the UAS's brownfield strategy.

Once the land is ready for production, UAS participants will have to confront the challenges posed by the market. In the light of the dominance of large corporations in the food production sector, it will be difficult for small, local farms and value-added enterprises to compete dollar for dollar with the agribusinesses that largely control the shelf space at most supermarkets. DSNI is going to have to reach out to and educate not only Roxbury consumers but those from the Boston area generally to create a market for locally grown, organic produce and value-added products. This will require a sophisticated marketing strategy and pricing structure to make UAS products attractive to area consumers.

Moreover, DSNI will have to convince the state and the City of Boston that Roxbury is a viable area for food production and related enterprises. Recently the Massachusetts Department of Food and Agriculture (DFA) announced a plan to develop a year-round, indoor public market in Boston for locally grown produce.[21] Harking back to Boston's last indoor public market, established at Faneuil Hall in the eighteenth century, the new market would offer local produce, baked goods, and fish, as well as local arts and crafts. Moreover, the market would contribute to the public life and public space of the city.

In the light of the UAS, Roxbury's agricultural heritage, and the amount of vacant, contaminated land in the neighborhood, the DSNI area would appear to be a perfect setting for the market. However, the DFA is eyeing a location for the proposed 40,000-square-foot market on the South Boston waterfront, to take advantage of its access to the seafood industry and Boston Harbor. The DFA and Boston planning officials have not seriously considered Roxbury as a potential site for the market. Just as DSNI must inevitably compete with agribusiness in the open market, it must also compete with public perception concerning the safety and desirability of Roxbury as a venue for new business ventures, including public projects. Like most other inner-city neighborhoods, the area has historically struggled to attract quality investment in the form of new businesses, especially retail supermarkets. The indoor public market could be a catalyst in developing the UAS and raising Roxbury's profile as an agricultural center. Nevertheless, decades of neglect and decay at the hands of the market stand in the way. The same forces that left Roxbury with a legacy of brownfields also conspire to repel potentially pathbreaking economic and social opportunities.

The case of the public market, like every other challenge confronting the UAS, suggests that the key to rebuilding Roxbury's damaged physical and social fabric is the community's ability to rally around its assets, such as land, and to promote them aggressively in the face of what is often an indifferent, if not hostile, public and private sector. To clean up its brownfield sites and attract investment in the UAS, DSNI must promote itself as a dynamic, diverse community uniquely positioned to create a model sustainable economy based on environmental improvement. With the blueprint for the UAS in hand, DSNI and the community as a whole are poised to reclaim Roxbury's place as a vibrant agricultural center.

5

Oakland's Fruitvale Transit Village: Building an Environmentally Sound Vehicle for Neighborhood Revitalization

There's no there there.

—Gertrude Stein on her birthplace, Oakland, California, in *Everybody's Autobiography*

People can't sit around waiting for someone else to do something for them. They have to do it themselves. The goal is to enable the residents of Fruitvale to make decisions about what's happening here—not external forces.

—Hayley Bryant, in the *Oakland Tribune*

Environmental Quality, Community, and Transportation in the Bay Area

There is perhaps no more beautiful metropolitan area in the United States than the Bay Area in northern California, where the glistening azure waters of the San Francisco Bay, the nation's largest inland bay, are encircled by the majestic copper spans of the Golden Gate Bridge to the west, the steadfast Marin headlands and inland redwood groves to the north, the steep Berkeley and Oakland Hills to the east, and the snow-capped Santa Cruz Mountains to the south.

Spanning 7,400 square miles, 9 counties, 100 cities, and over 6 million people, the Bay Area's built and natural attributes, plus a remarkably mild and sunny climate year round, have long conspired to make the area especially attractive for residents and visitors alike, offering a quality of life second to none.

Known as Yerba Buena until 1847, San Francisco started out as a mining town and emporium in the mid-1800s, bustling with the flow of goods up and down the coast, across the Pacific, and inland to the mines of the Sierra Nevada. It was an "instant city," built up almost overnight, and like so many other frontier cities, its inhabitants were driven by the pursuit of material gain afforded by the area's seemingly limitless natural resources, from vast stretches of undeveloped land, to fisheries, to gold and other valuable minerals discovered in the nearby Sierra foothills. In a mere forty years, San Francisco grew to a size that Boston required two and a half centuries to attain.[1]

Due east across the bay from San Francisco lies the city of Oakland. Its rise as the region's second largest city, less dramatic than its neighbor to the west, Oakland was first inhabited by the Ohlone Indian tribe and later became a vast land holding of the Mexican general Peralta into the nineteenth century. The city began as a series of farms and estates established by squatters, who eventually got title to most of Peralta's property by the mid-1800s. Not as prone to the fog that engulfs the San Francisco peninsula to the west in the warmer months of summer and early fall, the flat plains and undulating hills of Oakland and the East Bay afforded settlers sunny planting grounds for orchards of all kinds. With local access to the Western and Southern Pacific Railroads, Oakland farmers could transport their produce to markets well beyond the East Bay.

This merchantilist prowess spawned the Fruitvale district. Known as Brooklyn until the late 1800s, Fruitvale was the seat and commercial center of Alameda County until it was annexed by the city of Oakland in 1909. Located in the flatlands south of Oakland's central business district, Fruitvale was settled primarily by German immigrants and was long considered Oakland's second downtown because of its vibrant business and civic culture. The district attracted major retail establishments in the 1920s, such as Montgomery Ward, and developed a significant manufacturing base centered around the many canneries that served local orchards. With World War II came an economic boom and an influx of war industry workers, including large numbers of African Americans and Hispanics.

Following the war, the suburbs around the Bay Area burgeoned, owing largely to the freeway projects and mortgage subsidies that enticed middle-class urban residents to leave the increasingly crowded and polluted central cities of San Francisco and Oakland for the towns and subdivisions farther out. Meanwhile, urban neighborhoods like Fruitvale fell on hard times. Wartime factories closed, and workers were laid off. The canneries that once employed local residents began leaving the area in the 1960s.

In 1970, Fruitvale was still a majority white community; by 1980, it was majority black; by 1990, Latinos joined blacks as the dominant racial and ethnic groups, comprising almost two-thirds of the population, followed by Asians, whites, and a substantial Native American population. With each decade, the neighborhood experienced greater disinvestment and white flight. Jobs, housing, tax revenues, and other essential ingredients of community life dwindled, displaced to the suburbs to Oakland's east and south. Today more than 20 percent of Fruitvale families live below the poverty line, and almost a third of the residents are under the age of eighteen.[2]

Over the past three decades, urban sprawl resulting from the abandonment of central cities in the Bay Area has eaten away at the region's social and environmental stability from the inside out. In California as a whole, sprawl has resulted in a shortage of affordable housing, inaccessible jobs, and notorious traffic jams, leading to some of the nation's worst air pollution hot spots. In 1995, California-based Bank of America issued a report entitled *Beyond Sprawl* that declared, "As we approach the 21st century, it is clear that sprawl has created enormous costs that California can no longer afford."[3]

The primary cause of much of the Bay Area's sprawl has been the automobile. As with the rest of the country, suburbanization and the withering away of central city neighborhoods have been made possible by the development of massive roadways and the proliferation of cars that began in the 1950s. Transportation policy in the Bay Area, as elsewhere, has for decades favored cars over alternative forms of transportation such as mass transit, walking, and bicycling. As suburban commuters sit in traffic, lurching day in and day out toward the workplace or home on expensive highways paid for by public dollars, residents of inner-city neighborhoods contend with the noise, air pollution, and pedestrian hazards associated with that traffic, often without access to quality mass transit service or other offsetting amenities.

In Oakland, several major freeways, including Interstates 580, 680, 880, and 980, and hundreds of heavily traveled, high-capacity roadways pass through and above neighborhoods like Fruitvale. Motor vehicles going fast generate significant noise. For example, from sixteen yards away, a car traveling at 56 miles per hour makes ten times as much noise as it would going

31 miles per hour.[4] Noise from vehicular traffic typically drowns out all other ambient sound in congested urban neighborhoods and contributes to the same sense of congestion and unease that drivers themselves experience.

Further, the air pollution caused by cars, trucks, and buses driving through urban neighborhoods contains a deadly mixture of gases and particles. Transportation emissions, which include vehicle exhaust, fuel and paint vapors from gas stations and auto body shops, and tire dust, affect human health, especially in densely settled urban neighborhoods, where emissions tend to occur close to where people live and work. Also, because of frequent idling and stop-and-go traffic, urban transportation produces considerably more pollution per mile than does smooth-flowing traffic, when engines operate most efficiently. A lack of air pollution sinks or filters such as street trees and other vegetation can exacerbate the adverse effects of air pollution from motor vehicles.

People of color, who live in cities to a far greater extent than whites, are disproportionately exposed to urban air pollution. According to a California study, people of color are nearly three times as likely as whites to be exposed to harmful fine-particulate matter, emitted by sources that burn fossil fuels. Fine particles lodge in the lungs and, because of their size, cannot be expunged, resulting in respiratory and cardiovascular disease. The California study, which relied on readings from 161 air pollution monitors in different communities, found that 54 percent of the monitors in communities of color showed readings above the 1997 U.S. EPA health standard for fine particles. Only 19 percent of the air monitors in white communities had readings above the standard.[5]

In addition to particulate matter, motor vehicles that run on gas or diesel emit carbon monoxide and carbon dioxide, nitrogen oxides, and volatile organic compounds, resulting in environmental hazards such as ozone smog and greenhouse gases that cause global climate change. Transportation sources account for roughly one-third of greenhouse gas emissions. On the ground, these pollutants contribute to asthma, a serious and sometimes fatal illness that especially affects children. Asthma rates among children have risen 75 percent between 1982 and 1994, and young African Americans are four to six times more likely to die of asthma than are young whites.[6]

Beyond noise and air pollution, motor vehicles kill and injure urban pedestrians at alarming rates. In Boston, New York City, and San Francisco, half of the people killed in car accidents are pedestrians. Most at risk are children and the elderly, who depend most heavily on walking and bicycling for transportation.

Transportation infrastructure such as highways also usually comes at the expense of accessible open space for urban residents. Multilane highways invariably separate people from places like waterfronts or parks. Whether elevated or at groundlevel, such highways form impenetrable barriers of concrete, steel, and asphalt, shunting people from the precious few natural areas that remain in most cities. In Oakland, high-capacity surface roadways, including several interstates, criss-cross the city; the massive I-880 stands between the bulk of the city's residential areas and the Oakland waterfront, historically an industrial corridor that, in the Fruitvale and San Antonio sections, abuts a sleepy estuary separating the city from Alameda Island.

Urban residents have borne the brunt of the environmental harms associated with transportation, and most have not enjoyed access to quality mass transit services. Between the late 1960s and early 1990s, the average annual number of vehicle miles traveled by Americans doubled, reflecting increased suburbanization and highway construction and a concomitant decline in mass transit services. Yet in cities like Oakland, mass transit remains the preferred, if not sole, means of getting around for most people of color. Blacks and Hispanics account for almost half of transit riders in the United States, are four times as likely as Americans in general to take public transit to work, and are one and half times as likely to walk to work.[7] Access to affordable, reliable transit services means access to jobs, shopping centers, parks, and other basic building blocks of a healthy community. In Oakland, the Church Community Jobs Commission has tried to help local African Americans gain employment with Bay Area employers such as AT&T. However, because most jobs are located in distant suburban communities, Oakland job seekers who do not own cars have no way of getting to work.[8]

Notwithstanding the importance of mass transit to urban communities of color, most state and local governments have failed to invest in transit infrastructure improvements. Instead, they have subsidized projects that serve suburban commuters, such as highways and commuter rail lines. Until recently in Los Angeles, for example, the Metropolitan Transit Authority

(MTA) spent 70 percent of its discretionary funds on a rail system that served only 6 percent of the passengers; the MTA's buses, serving 94 percent of the passengers (many of whom are lower income and 81 percent of whom are people of color), got only 30 percent of the funds.[9]

In Oakland, residents of the city's West Side battled Caltrans, the state highway agency, to stop the rebuilding of the Cypress Freeway after it was damaged by the Loma Pieta earthquake in 1993. The Church of the Living God Tabernacle, along with the Clean Air Alternative Coalition, claimed that "freeways have caused high cancer rates in the communities alongside them and high incidences of lead in the brains of our children living in these communities." The area abutting the freeway is 92 percent people of color.[10] The community groups filed a civil rights lawsuit, alleging that the proposed freeway reconstruction violated Title VI of the Civil Rights Act of 1964, which prohibits discrimination in any federally funded project. "The brunt of the proposed project's negative social, human health and environmental impacts," the groups argued, "including those associated with noise and air pollution, the dislocation of persons, the condemnation of homes and businesses, the chilling of economic development, as well as the disruption of the life of the community—will be borne by minority residents of West Oakland."[11]

That the residents of Oakland's West Side had to resort to civil rights laws to protect their environment underscores the sometimes pernicious limits of traditional environmental regulations. The unchecked development of massive highway projects, proliferation of cars, and lack of access to open space and the Oakland waterfront demonstrate that the urban environment has been the orphan of traditional environmental law and policy. Although California's air quality regulations are the strictest in the nation due to the state's unique and persistent air pollution problems, the environmental impacts of highway projects meant to serve regional commuters have traditionally been reviewed according to regional air quality conditions, not local hot spots. As in most other jurisdictions, California's health-based regulations look at air quality on a parts-per-million, pollutant-by-pollutant basis and tend to discount the cumulative, compounded health and environmental effects of many different kinds of pollutants in the atmosphere, especially as they affect people living or working close to the pollution sources.

Moreover, air pollution from mobile sources such as cars is regulated with standards that dictate how much pollution a single car can emit from its tailpipe and how many miles per gallon the car's engine must achieve. California's air quality regulations, like those in every other state, say nothing about how many cars can be produced or driven or how many highways can be built to accommodate them. Thus, air quality gains from tougher emissions and efficiency standards can be wiped away by the proliferation of more and more automobiles riding on ever-widening roadways.

Meanwhile, the California Environmental Quality Act, the state's environmental impact review statute, examines projects in isolation, one at a time. Each highway development in Oakland has been dealt with on its own terms and ultimately has been driven by the need for greater roadway capacity to handle the regular increases in Bay Area commuter traffic. This piecemeal review has resulted in both increased air pollution from highway traffic and the isolation of most Oakland residents from the waterfront and the loss of open space. With little in the way of comprehensive land use planning or growth management in Bay Area communities, the neighborhoods in Oakland's flats have been left to deal with the adverse environmental consequences.

As civil rights leader and U.S. congressman John Lewis explains, "Even today, some of our transportation policies and practices destroy stable neighborhoods, isolate and segregate our citizens in deteriorating neighborhoods, and fail to provide access to jobs and economic growth centers."[12] Transportation is the vehicle by which a community develops; it is a critical enabling mechanism that allows individuals within a community to engage in professional, recreational, cultural, and other essential activities. According to attorneys Steve Burrington and Bennet Heart of the Boston-based Conservation Law Foundation,

> Each choice [concerning transportation] will have a major impact on the neighborhood, and on the lives of people who live there: There will be easy access, or poor access, or no access at all, to jobs in other places; One day next year, a third grader walking to school will get there safely, or be hit and killed by a car; Noise levels will be reasonable, or a serious source of stress; There will be high levels of diesel exhaust and other pollution in the air, or pollution levels that don't harm human health; Neighbors will spend time on the sidewalk talking to each other, or will avoid the roar of traffic, stay inside and let criminal activity take over; People from other neighborhoods will come to patronize local businesses, or vacant storefronts will multiply.[13]

Transportation thus exercises a profound influence on the quality of life of urban neighborhoods. It is a critical part of the fiber, the infrastructure, that holds urban communities together and enables urban denizens to pursue the variety of daily activities that make up their lives and livelihoods. As author Jane Holtz Kay describes, "A . . . community and transit are intertwined: to secure transit service, you need community; to secure community, you need transit."[14]

It seems only natural that efforts to revitalize a hard-hit urban neighborhood might start with transportation as a means of bringing people together, literally and figuratively, and helping to restore a safe, healthy environment. In Oakland's Fruitvale neighborhood, where residents have seen the worst of what automobiles, pollution, and disinvestment can do to a community, that is exactly what is happening.

The Road to Recovery: Fruitvale's Transit Village

Along Fruitvale Avenue, the crowded boulevard that runs southwest from the Oakland hills toward the San Francisco Bay, forming the northern border of the Fruitvale district, the offices of the Unity Council (UC, formerly the Spanish-Speaking Unity Council) bustle with activity. Formed in 1964 by activists seeking to create a forum to address issues of concern to the growing Latino community, UC began as a civil rights organization, later evolving into a community development corporation focused exclusively on Fruitvale.[15]

UC's director, Arabella Martinez, began her career as an Alameda County welfare case worker after completing graduate school at the University of California, Berkeley. She was among the group of activists who founded UC in 1964. Her commitment to the community stems from her sense of place. "I certainly have a sense that this is my community," she explains, "I've always had a place here. I'm rooted here."[16]

Under Martinez's leadership, UC designed a comprehensive approach to dealing with Fruitvale's challenges that emphasizes the neighborhood's strengths and assets rather than its weaknesses. Focusing on Fruitvale's rich tradition of commercial activity and ethnic diversity, UC has been able to bring together residents, community organizations, and businesses in delivering successful community projects such as affordable housing, child care and senior programs, and community centers. As well, unlike many other lower-income neighborhoods, Fruitvale has been able to maintain a vibrant business district along the neighborhood's main commercial thoroughfare, East Fourteenth Street, considered a regional destination for Latino shoppers. The Fruitvale Community Collaborative, a consortium of local groups spearheaded by UC in 1991, helps organize local residents and businesses in establishing block associations and conducting community cleanups and tree plantings, traffic-calming campaigns, and flea markets.

Yet as is so often the case in hard-hit urban neighborhoods, no matter how well organized or successful, community residents and organizations must constantly contend with unwanted development. A case in point was a 1991 proposal by the Bay Area Rapid Transit authority (BART) to build a 500-car parking facility at its Fruitvale station, in the middle of the area's residential and commercial center.

Environmental issues had always weighed heavily on the minds of Martinez, her UC colleagues, and the Fruitvale community as a whole. Restoration of Sausal Creek, one of the few vestiges of interior Oakland's connection to its waterfront, and cleanup of hazardous waste sites in the neighborhood have been on UC's action agenda for years. Yet the neighborhood's greatest environmental challenge has been local traffic conditions and the poor air quality those conditions have caused. Air pollution, noise, and pedestrian hazards have long plagued Fruitvale residents, with little sign of improvement.

When BART officials approached the neighborhood in 1991 with plans for a multilevel concrete parking structure on a parcel abutting the station, UC and neighborhood residents came out in force against the proposal, arguing that the facility, a massive eyesore, would not only exacerbate existing air quality and traffic problems but would act as a dangerous barrier, severing pedestrian access to the East Fourteenth Street business district. "We opposed it," Martinez says, "because it would have further separated the BART station from the neighborhood."[17] Moreover, UC claimed that BART had failed to involve the community in its planning and decision making. Eventually BART dropped the proposal, citing concerns about potential adverse air quality impacts due to increased traffic.

Notwithstanding its successful opposition to the BART proposal, UC recognized that the BART station itself was a significant neighborhood asset, providing a convenient, low-cost alternative for the many area residents who do not own cars and making the neighborhood accessible to residents from other Bay Area communities. In 1991, Rich Bell, a graduate planning student at Berkeley, working with Martinez and UC staff, saw the parking garage proposal as an opportunity for UC to present its own ideas for development of the Fruitvale BART station, which would take advantage of the station as a neighborhood asset. In his courses at Berkeley, Bell had learned about the new planning concept called "transit-oriented development," which sees mass transit as a lever for neighborhood revitalization and environmental improvement. Transit-oriented development attracts ridership (thus reducing traffic and air quality impacts) not by expanding parking infrastructure on site (which merely increases vehicle miles traveled through the neighborhood), but enhancing existing amenities and increasing residential development near transit stations. Traditionally, BART emphasized parking capacity as the key to attracting riders. As a result, most of BART's thirty-four stations, like Fruitvale, are surrounded by sprawling surface parking lots.

At the heart of the new approach is the concentration of mixed-use development—high-density housing, shops, and public spaces—within a quarter-mile radius of transit stations. Concomitantly, transit-oriented development discourages density development elsewhere in the area. In essence, development is driven by pedestrian access to transit, which makes transit more appealing to riders.

In California, the push toward transit-oriented development was driven largely by the state's stringent air quality regulations, which require sharp reductions in pollutants like carbon dioxide emitted by automobiles. Research has shown that transit-oriented development can reduce auto trips by as much as 18 percent, and several reports have demonstrated dramatically higher BART ridership among people living within a quarter-mile of transit stations than among the general public. Moreover, these commuters tend to walk to the station, thus avoiding the automobile engine cold start, a major contributor to regional air pollution. Also, commuters driving one-half mile to BART stations cause almost the same amount of pollution as commuters who drive ten or fifteen miles to work.[18]

Using this new policy concept, Martinez, Bell, and UC were able to transform their opposition to the BART parking garage into a sophisticated plan for community-based economic revitalization, enlisting BART officials in their effort as codeveloper with the Fruitvale Development Corporation, UC's economic development wing. "This [was] not the usual planning process," Martinez recalls, "It came from the community and the people that live here. That's new. . . . By working with us and meeting the needs of the community, the costs of development [would] be much less because we [wouldn't] be fighting all the time. We really do want development here, but we want it done in terms of what this neighborhood will support."[19]

Martinez and her colleagues reckoned that most new housing, office, and retail space is built on the fringes of cities, in suburbs. They understood that this kind of development results in region-wide environmental and economic problems. For example, suburban development destroys prime farmland, open space, and plant and wildlife habitat, and it causes traffic congestion and air pollution. In addition, it draws investment away from central cities, further distancing inner-city residents from much-needed jobs and economic opportunities. For BART officials, UC's transit-oriented approach was just what they were looking for. "Fruitvale for 20 years has had various programs for revitalization, none of which has gone anywhere," explains BART director Michael Bernick, "This is the first program that tries to use the transit station both as the center of and the spur to development. . . . This is a very important step."[20] Before Fruitvale, every large-scale transit-oriented development project undertaken in the United States had been located in a new development or wealthy suburb. This would be the nation's first such project in a lower-income, inner-city community.

Immediately UC began to engage community residents in a visioning and planning process to flesh out the parameters of what was soon being called the Fruitvale Transit Village. Oakland's mayor in 1992, Elihu Harris, having been elected partly on his promise of changing the focus of city government from downtown to neighborhoods like Fruitvale and responding to a strong turnout by Fruitvale voters, supported the effort by earmarking $185,000 in city Community Development Block Grant funds to jump-start the planning process. UC raised an additional $290,000 from government and foundation sources, including the federal Intersurface Transportation Equity Act (ISTEA).

Beginning in 1992, UC acted as a facilitator and convener, bringing together different stakeholder groups from around the community and disseminating the results of the planning process. In the spring of 1993, working with the planning department from Berkeley, UC sponsored a community design symposium, or charette, to review and comment on various design proposals for the transit village. UC then held a series of community planning meetings, based on feedback from the charette, to begin to develop a comprehensive redevelopment plan.

The village would be located on the current BART parking lot, a nine-acre site bounded by Fruitvale and Thirty-Seventh avenues on the west and east, respectively, and East Twelfth Street and the BART station on the north and south, respectively. The lot accommodates over a thousand vehicles. As required by BART policy, replacement parking would have to be integrated into the project's design so as to maintain existing capacity. In addition to the environmental benefits that would result from increased ridership, early planning discussions embraced a vision of a transit village where pedestrians and bicyclists could enjoy easy access to local businesses, a health care clinic, a child care facility and community center, UC offices, affordable housing, and even a Fruitvale museum showcasing the area's Latino heritage. The center of the village would be a capacious, elegant pedestrian plaza, lined by trees and filled with people around the clock.

By the spring of 1993, the community planning effort began to receive national attention. In April, U.S. Transportation Secretary Federico Peña visited the Fruitvale BART station, promising to push for federal aid for the project. He returned in August to report that the transit village planning effort would receive a $470,000 grant from the Federal Transit Administration. By this stage in the planning process, money had become a critical factor. As the scale of the project unfolded, each month grander than the last, the need for funding to match the vision increased. In the summer of 1995, as UC convened a series of community planning meetings to develop final project goals and an overall design, and prepared to enter the technical phase of the development process, including environmental review, traffic studies, and economic modeling, the projected cost of the transit village had reached tens of millions of dollars.

Hoping to receive $20 million in federal Empowerment Zone funds, UC learned in October 1995 that it would get only a fraction of that amount: $3.3 million in grants and $3.3 million in loans. Also in October, BART officials informed UC that money it expected BART to contribute to the project as a result of the aborted parking garage, $15 million, would likely not be made available. Suddenly the need for dollars was driving the dream, diminishing and reshaping it.

In an effort to raise the funds necessary to make the Fruitvale Transit Village a reality, UC hired Chris Hudson, an urban planner by training, as project manager in the late summer of 1995. As Hudson explains, his job "was to make the money match the vision."[21] Undaunted by October's setbacks, Hudson, Martinez, and their UC colleagues geared up for some creative fund raising, and not a moment too soon, for the project was moving from the planning phase into action, with environmental review about to begin.

Because the transit village would be developed on what was once industrial land, hosting an assortment of uses, from a vulcanized rubber plant and electroplating facility, to an auto body repair shop and railroad spur, UC had to deal with the costs and potential health risks posed by brownfield cleanup. With some leftover planning money and a grant from the U.S. EPA, UC was able to conduct a comprehensive site assessment and design a cost-effective and safe cleanup strategy. As Hudson explains, UC "quickly became experts in environmental cleanup. We tried to take a creative approach to dealing with soil and groundwater contamination, recognizing that we couldn't afford to do everything. We feared potential liability, but were confident that between the level of cleanup we're doing, and the project we're building, environmental conditions will be greatly improved."[22]

During the years 1996–1997, the transit village took on a definitive shape and scale. After over three years of community involvement in planning and design, the community, led by UC, embraced the transit village as the key to transforming Fruitvale's commercial and transportation core into a vital, healthy place to live, work, and shop. As Hudson explains, the transit village will revitalize the area "because it builds upon two of Fruitvale's greatest assets: accessibility to transit, and a strong network of community-based organizations. The image of the neighborhood will be enhanced and reinvestment will be catalyzed, creating a safe, attractive, and pedestrian-oriented commercial center."[23]

The heart of the project is a pedestrian plaza adjacent to the existing BART station entrance, lined with small shops and restaurants. It will be a venue for neighborhood festivals and concerts. Vendors will sell food and other goods from pushcarts, while neighbors commune on tree-shaded benches. Surrounding the pedestrian plaza will be approximately 146,000 square feet of new space for commercial and nonprofit use. There will be a health care clinic, La Clinica de la Raza, the Fruitvale–San Antonio Senior Center, a day care center, a branch of the Oakland Public Library, and offices for UC and other community service providers. Also near the plaza will be close to 18,000 square feet of existing retail space. Additionally, the transit village will have approximately fifteen residential units. Three hundred twenty-five new parking spaces will be created, as well as a parking structure for relocation of 979 existing BART surface parking spaces. Finally, East Twelfth Street will be narrowed from two lanes to one in each direction.

The design of the transit village reflects the core goals and principles underlying UC's revitalization efforts:

• *Strengthening existing community institutions.* The transit village will allow community members and institutions to work together more easily and efficiently by housing them in a single, convenient location. UC hopes to build the civic will and capacity of Fruitvale by creating a vital physical and social nerve center for the community.

• *Providing a stable source of jobs and income for the community.* Currently 90 percent of Fruitvale's workers have to go outside the neighborhood for work. With new nonprofit and commercial ventures, the transit village will be a source of jobs, for which workers will not need to own a car. UC estimates the transit village will create over seven hundred new jobs. As well, UC hopes to incubate many small businesses and microenterprises within a public marketplace on site, providing technical assistance and preserving the locally owned, small business flavor of Fruitvale. The proximity of the BART station will provide a large base of potential customers for local businesses, while attracting new private investment.

• *Increasing the variety of retail goods and services available in the community.* As with employment opportunities, many local residents frequently leave the neighborhood to buy many kinds of goods and services. By broadening the types of retail shops and services available, local income and taxes will remain in the community.

• *Beautifying a blighted area.* The transit village will convert sprawling parking lots, brownfield sites, and traffic-congested streets into an attractive community and commercial space centered around a pedestrian plaza.

With street and facade improvements as part of the development, the transit village will serve as a catalyst for community-wide beautification.

- *Increasing real and perceived safety.* Like so many urban other neighborhoods, where blighted lots and abandoned buildings dominate the viewscape, the area around the Fruitvale BART station is widely considered a magnet for criminal activities. With housing, commerce, and a steady stream of pedestrians, the transit village will ensure that crime is minimized, if not eradicated altogether. From improved lighting to traffic-calming measures, the transit village will reduce the opportunities for crime and demonstrate that Fruitvale residents care about and are invested in their neighborhood
- *Providing high-quality affordable housing.* As the quality of life in a neighborhood improves, invariably it becomes more expensive to live in. Already, Oakland and the Bay Area generally are among the nation's most expensive housing markets. The transit village will alleviate some of this pressure by providing lower-income residents affordable rental housing, with access to nearby jobs, shops, and other community resources.
- *Encouraging and leveraging public and private investment.* Because the transit village will improve the look and sense of the neighborhood, highlighting its resources and assets, it will attract outside investment and encourage local home owners and businesses to invest in improvements.
- *Increasing BART ridership and reducing traffic and pollution.* The genesis and core of the transit village lies in the notion that, by reducing the number of cars on neighborhood streets, the neighborhood will be made safer, cleaner, quieter, and healthier. Environmental improvement thus becomes the catalyst for community-based problem solving addressing the full range of neighborhood concerns, from jobs and housing to health care and crime.

As the final plans came together in 1997 and early 1998, the project was subject to intensive environmental review under the National Environmental Policy Act and the California Environmental Quality Act. Because UC and BART believed the project would improve, not degrade, environmental conditions in Fruitvale, they asked for and were eventually granted in May 1998 a "mitigated negative declaration" from Oakland's Planning Commission, determining that the project would not have a significant impact on the environment because of certain mitigation measures incorporated into the project, including cleanup and disposal of hazardous materials in the soil, replacement parking for BART riders, and proper signage and crossings for bicyclists and pedestrians. To ensure that the project would not adversely affect traffic and air quality conditions in the neighborhood, UC petitioned the city of Oakland for a zoning ordinance to prohibit additional parking spaces in the area around the transit village. In essence, UC wanted to codify its commitment, and that of the community, to forms of transportation other than cars. In the fall of 1996, the zoning ordinance became law.

With the project scaled down somewhat from earlier plans (for instance, no museum and fewer housing units), UC was able to secure funding in the fall of 1997 to develop the replacement parking facility that was critical to building the rest of the project; without it, BART could not allow the project to proceed. Subsequently, additional funding commitments came through from private foundations and public sources, enabling UC to move forward with its predevelopment activities. The groundbreaking took place in September 1999, and the project is on schedule to be completed by the end of 2000.

Going One Step Further: Restoring Public Access to Oakland's Waterfront

Because of the momentum and visibility of the transit village project, UC has been able to leverage its success to benefit other efforts aimed at improving the quality of life in the neighborhood. For example, UC has raised funds to help local businesses not directly associated with the transit village undertake facade improvements as part of the National Trust for Historic Preservation's Main Street program. According to Chris Hudson, "The transit village project has provided an opportunity to look at the community as a whole and to see the connections. . . . The project is a catalyst for positive change in ways we might not have anticipated going into it."[24]

Feeding off the civic energy created by the transit village effort, UC and several community partners formed the Fruitvale Recreation and Open Space Initiative (FROSI) in early 1997. Working with Oakland's Office of Parks, Recreation and Cultural Services, the University-Oakland Metropolitan Forum (a community development partnership between the University of California at Berkeley and the city of Oakland), the California State Coastal Conservancy, the Friends of the University of California at Berkeley Crew Team, and the Trust for Public Land, UC and community residents are trying to increase Fruitvale's and the adjacent San Antonio neighborhood's environmental assets by converting a degraded nine-acre vacant lot on Oakland's waterfront into a flagship community park, called Union Point Park.

Michael Rios, UC's FROSI project manager, explains that FROSI takes an entrepreneurial, asset-based approach to the natural and recreational resources of the area. As with the BART station, UC and community residents recognized that the Oakland waterfront was a highly underused and undervalued community asset. With nineteen miles of shoreline, Oakland has one of the longest Bay Area shorelines, yet few people think of Oakland as a waterfront city. Historically, the Oakland estuary has been inaccessible to the public because of a host of industrial and military facilities located on the water's edge. "Oakland and the people that live here have felt cut off from the waterfront. . . . Right now, the waterfront is not a very friendly place to go to. You don't want to get out of your car in some places," says Virginia Hamrick, a member of the Oakland Estuary Advisory Committee, the group charged with implementing the Oakland Estuary Plan, a long-range planning effort to determine the fate of Oakland's waterfront.[25] FROSI is a part of the advisory committee and serves as a direct link to community residents.

Moreover, according to Rios, in spite of having the highest proportion of children in Oakland, Fruitvale and San Antonio have the lowest proportion of parks and open space.[26] Oakland's General Plan, the city's master planning document, established a city-wide standard of four acres of park land for every one thousand residents. In Fruitvale and San Antonio, the average is less than one-half acre per one thousand residents.

With the city and the Port of Oakland as partners, FROSI has engaged in a bottom-up planning process in which area residents are participating in the decision making concerning the park's design. Through on-site planning sessions, including a special Earth Day event in April 1998 and a workshop with area youth held the following September, FROSI has developed a comprehensive plan for Union Point Park.

As proposed, the $4 million park will provide diverse recreational opportunities, including boating, a sports field, and picnic facilities, and it will host the Berkeley crew team's boat house. With the removal of abandoned industrial buildings and other hazards that prevent access to the waterfront, the park will complete a section of the San Francisco Bay Trail, connecting already-completed sections, while serving as a staging area for the trail system. Through the planting of trees and vegetation and shoreline restoration, the park will provide habitat for shorebirds and other wildlife.

Aside from environmental and recreational benefits, FROSI sees the park as an opportunity to contribute to the area's overall economic development and revitalization. With close to $1 million already raised for the acquisition of privately owned land and design and predevelopment costs, UC and its FROSI partners are well on their way to making Union Point Park a reality.

Civic Environmentalism and the Fruitvale Transit Village

In a 1996 report, *Blueprint for a Sustainable Bay Area*, the Bay Area environmental group Urban Ecology spelled out the challenge that now confronts Bay Area communities: "The San Francisco Bay Area has reached a pivotal point in its history. . . . Because of past patterns of planning and development the region is losing the special qualities that have made it a desirable place to live and work for generations. The time to start changing these destructive patterns and chart a new course for the future is now."[27]

The Fruitvale Transit Village and Union Point Park represent a powerful civic environmental response to this challenge. UC's commitment to participatory planning and a proactive, asset-based approach to community development and environmental protection, defines the transit village and park projects and accounts for their success to date. Notwithstanding the projects' considerable financial demands, starting in 1991 and continuing to the present, UC has been able to enlist a broad range of funders, partners, and champions because of its innovative, affirmative strategy. As well, UC has been able to integrate professional planners, architects, and developers into its bottom-up, citizen-led initiative. Martinez and her UC colleagues have effectively balanced the technical, professional aspects of community development with their commitment to the democratic values of participatory decision making and accountability.

Despite occasional setbacks, resulting in the scaling back of the original transit village design, UC's ability to engage community residents, professionals, and BART in a long-term, iterative planning process proved decisive. Recognizing in local air quality and traffic problems an opportunity for neighborhood improvement and exploiting the state's push for increased mass transit ridership, UC took the high road in opposing the 1991 proposed BART parking facility and became, in effect, the lead agency, a term usually reserved for government actors, in promoting the transit village concept.

Thus, what was at first merely another case of environmental "us versus them" was transformed into a dynamic, collaborative process that went well beyond environmental and land use issues to incorporate a community-wide revitalization strategy. As evidenced by the Union Point Park project, the momentum generated by the transit village spawned new opportunities for change, which created exciting synergies between the transit village project and other neighborhood development initiatives.

Moreover, as both Chris Hudson and Michael Rios emphasize, consistent community involvement allowed a large, educated constituency to develop, who understood the connection between environmental improvements and overall community change and who were prepared to capitalize on opportunities for action. Notwithstanding the flaws in existing environmental laws, which had allowed neighborhood conditions to deteriorate over time, UC in effect moved beyond law in developing its comprehensive strategy while still working within the parameters of environmental cleanup and review standards.

Fruitvale residents, most of them lower-income people of color, carried the banner of environmental justice in their campaign to improve transportation and overall environmental quality. Having put up with traffic, air pollution, and lack of access to quality parks and open space for decades, UC and the Fruitvale community took a stand, and in the process transformed their environmental problems into a strategy for economic, social, and environmental revitalization.

Moreover, to the extent that communities of color, especially Latino communities (already 65 percent of the West's population), in California and throughout the West will continue to grow, Fruitvale's success to date is a welcome harbinger of a multicultural future, where sustainable communities are built by a diverse group of local citizens, organizations, businesses, and government agencies. UC's experiment in environmental action and community building appears to give the lie to those who, like author Robert Kaplan, predict only greater environmental and social decline in the twenty-first century, where sprawling suburbs, poisonous urban air, and segregated communities are the norm.[28]

Although not an explicit part of UC's approach, industrial ecology principles can be gleaned from the transit village and park projects. As Michael Rios explains, UC takes an ecological approach to community building in that it views the community as a system, each part intricately bound to the other. This approach allowed UC to develop the transit village idea in the first place, recognizing the connection between locally based economic development and environmental quality. The Fruitvale Transit Village and Union Point Park embody the idea of sustainability that is at the heart of industrial ecology. The environmental group Urban Ecology, of which Rios is a director, offers a compelling definition of sustainability implicit in UC's work: "A sustainable Bay Area supports and improves the quality of life of all its residents and recognizes the interdependence of its people, culture, economy, and urbanized and natural environments."[29] UC never conceived environmental protection as separate from economic or social issues; to Martinez, Hudson, Rios, and their colleagues, the community's physical conditions are part and parcel of its economic and social conditions. The environment and community are one.

The concept of place is also at the core of the transit village and park projects. Over time, Fruitvale residents lost touch with their physical surroundings owing to the assaults of traffic, highway infrastructure, and blighted land. The transit village and park are designed to restore a sense of place and security to the hard-hit Fruitvale district, to heal and make whole the community. In this sense, UC has embraced the poet Gary Snyder's description of the power of place: "Of all the memberships we identify ourselves by (racial, ethnic, sexual, national, class, age, religious, occupational) the one that is most forgotten, and that has the greatest potential for healing, is place."[30]

The jury is still out on just how successful the transit village will be in increasing the number of pedestrians and bicyclists while decreasing the amount of automobile traffic in the neighborhood. Further, a weak regional and national economy could undermine UC's ability to create a viable commercial center within the transit village. And if market forces continue to channel growth to outlying suburban areas, Fruitvale will continue to struggle in its attempt to improve local economic and environmental conditions. Yet in the light of the commitment of a diverse set of stakeholders to the projects' success, from community residents and nonprofit organizations like UC, to BART and other government agencies, to private foundations, it is likely that UC's efforts will continue to generate momentum and results. Like DSNI's Urban Agriculture Strategy, the Fruitvale Transit Village and Union Point Park signal a bold and promising approach to a new kind of urban renewal, in which the obstacles to community progress, from commuter parking lots and highways to brownfield sites, are now being leveraged as tools for revitalization.

6

Community-Based Conservation and Conservation-Based Development in Rural Colorado

It is a quick ride from the Geography of Hope to the cliff of fear.
—Timothy Egan, *Lasso the Wind*

Conservation . . . is all about who the people are.
—Lynne Sherrod, interview

Land Rush Redux: The New Flight to the Frontier in the Rocky Mountain West

The vast stretches of open country and rugged vistas of the Rocky Mountains evoke grandiose sentiments and national pride like no other part of the country. Freedom and opportunity, values at the core of our democratic faith, take shape in the valleys and mountains that stretch from Montana south along the Continental Divide to New Mexico. This is the kind of boundless territory that Jefferson believed would perpetuate a strong republic, the "empire for liberty," the pursuit of which compelled him to send Meriweather Lewis and William Clark on their famous journey up the Missouri River.[1]

Paradoxically, it was the eventual settlement of the West and these hardscrabble places—the frontier—in the late nineteenth century that prompted the historian Frederick Jackson Turner to warn of the impending decline of the American polity, fearing the loss of the frontier's democratizing effect. For Turner, the development of the West, beginning with the free land era and homesteaders in the 1850s and 1860s, and ending with the large-scale settlement of the Pacific Northwest in the 1890s, was cause for grave concern about the fate of the republic, not to mention the destiny of countless bands of Native Americans displaced or destroyed by western settlement, a matter Turner conveniently overlooked.

Yet despite Turner's anxiety and the cruel conquest of western tribes, the West has continued to hold out signs of democratic hope to those who have gone in search of them. For many who call the West home, the region represents the apex of nature and culture, celebrated by the famous eighteenth-century anglican bishop George Berkeley in his musings on the promise of the West, earning him the honor of having a city named after him—Berkeley, California. He wrote in *On the Prospect of Planting Arts and Learning in America:*

> Westward the course of empire takes its way
> the first four acts already past
> A fifth shall close the drama with the day
> Time's noblest offspring is the last.[2]

The sheer size of the Rocky Mountain states has afforded the unique opportunity to live close to nature in a way people cannot in the more crowded and urbanized areas of the East and Midwest. The eminent writer of the West, the late Wallace Stegner, summed up the feelings of the countless souls who have ventured beyond the Mississippi seeking a better life, or who have at least imagined doing so. Stegner wrote in *The Sound of Mountain Water,* "One cannot be pessimistic about the West. This is the native home of hope." For Stegner, the West's seemingly endless open country embodied the possibility and opportunity that endowed the region with its power and resilience. But Stegner also recognized "what heedless men have done and still do to [the] noble habitat" of the West, the environmental assaults that, for those less sanguine than Stegner, have finally begun to eclipse their indigenous optimism.[3]

In the second half of the twentieth century, cities of the Rocky Mountain West have experienced three major growth periods, resulting from the perennial allure of the promise of the good life and ready access to an arcadian setting. Denver and Salt Lake City are among the fifteen fastest-growing metropolitan areas in the nation, on average, growing at the rate of 50,000 people per year.[4] In the West as a whole, urban areas have ballooned. Sixty years ago, fewer than half of all westerners lived in metropolitan centers. Today 86 percent reside in cities, a higher percentage than in New York, at 80 percent.

The urbanization of the West, and of the Rocky Mountain region in particular, has led to unprecedented environmental degradation, transforming the rugged landscape and expansive views into yet another case of unchecked development. In Colorado, major roadway projects have allowed traffic in Denver to increase at twice the rate of local population growth, undermining the city's effort to clean up the dirty brown air that often hovers over the region. Meanwhile, low property tax rates have encouraged development that has consumed parks and open space at levels well beyond the population growth rate. Prohibited by state law from raising property taxes beyond a certain level, municipalities have been forced to generate most of their revenues from businesses, thus encouraging retail and commercial sprawl. Since the late 1980s, over 500,000 acres of open space have been lost to development across the state. New subdivisions and other development are chewing up farmland at the rate of 90,000 acres per year; every hour, 10 acres are being converted to sprawl.[5] As a result of this growth and development, many Coloradoans now face a water crisis: some new Denver suburbs, many predict, will run out of water within a decade.

Douglas County, Colorado, is a case of growth run amok. The nation's fastest-growing county, Douglas County was developed almost overnight on the prairie south of Denver. With nine new schools, the county still has nearly the lowest per student spending ratio in the state because Colorado's tax system limits the amount of money the school district can generate from property taxes.[6] Alongside Highway 36, between Denver and Boulder, tens of thousands of new homes and the headquarters of dozens of companies dominate the landscape. In Douglas County, journalist Timothy Egan describes the Rockies as "close enough to feel their breath. But on some days, the mountains are obscured by Denver's brown cloud, drifting north. At rush hour, Highway 36 is a parking lot all the way back to Denver."[7] Today the Denver metropolitan area stretches 535 square miles, or as far as the naked eye can see. By 2020, the Sierra Club warns in its landmark study, *The Dark Side of the American Dream: The Costs and Consequences of Suburban Sprawl*, that area could expand to 1,150 square miles.[8]

The problem of urban sprawl extends well beyond Denver's ever-expanding borders. In Routt County in northwestern Colorado, about three hours from Denver, or 150 miles as the crow flies, the search for the idyllic

Rocky Mountain lifestyle has begun to effect significant changes on the area's environment and communities. Seeking refuge from the congestion and pollution in cities like Denver, Salt Lake City, and beyond, many tourists and second-home buyers are threatening to transform the character and culture of rural Routt County, a majestic land of snow-capped mountains and rugged valleys perched high on the Colorado Plateau. With fifty full-time ranches and sixteen thousand head of cattle, Routt County maintains a strong agricultural tradition but is under assault. Increasingly, thirty-five-acre subdivisions, notoriously exempt from environmental review by Colorado Senate bill 35, are becoming the norm, slicing up the prairies and ranchlands that comprise the core of the county's identity and economy.

Not only in Colorado but throughout the rest of the Rocky Mountain West, counties once considered too rural and remote to attract much attention have become safe havens for many Americans who have grown weary of the traffic, smog, and crime of the cities and suburbs. From 1990 through 1995, rural counties across America saw a net influx of more than 1.6 million people, almost all of it from domestic migration and almost all of it white. In the Rocky Mountain states, most of the fastest-growing rural counties are at least 85 percent white, creating racial and ethnic balkanization.[9] Moreover, the rapid growth of rural areas and accompanying land rush have resulted in unwelcome changes in the culture and quality of life of rural communities. Real estate values and housing costs have escalated, unsightly strip development has altered the fragile landscape, productive agricultural land and other natural resources have been diminished, and the small town flavor and neighborliness of many communities have been undermined. In effect, urban sprawl has bred rural sprawl, resulting in a degenerative pattern of growth, decay, and flight.

The environmental and social changes brought about by growth and development in the Colorado countryside have given pause to many, who have begun to ask how this could happen to their beloved landscape and what can be done to stop it. Ray Studer, director of the Urban and Regional Planning Program at the University of Colorado at Boulder, explains that the sprawl and pollution affecting the region are in part a cultural problem. "Nobody in Colorado is willingly trying to screw up the land or dirty the air," Studer argues, "This kind of stuff happens because we Westerners are not, by nature, a planning culture."[10] Former Colorado governor Roy Romer also points to a cultural predisposition away from planning and toward private property rights as the reason efforts to contain growth and sprawl have failed. "It's an old Western ethic: don't tell me what to do on my property," Romer explains.[11] The physical transformation of Colorado, emanating from its cities to its rural areas, is but a symptom of underlying cultural and social forces resistant to system-oriented community planning (as in smart growth planning). The resultant deficit opens the door to negative environmental effects like urban smog, which contribute to white flight and racial balkanization, which in turn lead to loss of working landscapes and open space and the decline of traditional ways of life.

Because environmental laws were designed to control pollution on a pollutant-by-pollutant, source-by-source basis, and were not designed to deal with ecosystem-wide degradation, let alone the destruction of local and regional cultures, Coloradoans have had to resort to extraordinary measures to try to curtail the state's environmental and social decay. In 1996, Colorado voters passed a constitutional amendment allowing the state to designate certain areas as protected open space, precluding development. In a similar move, the Denver Regional Council of Governments recently adopted a bold long-range plan for managing the area's growth, calling for protection of prime agricultural and rural lands. In addition, the Colorado Responsible Growth Act, which directs growth to certain designated areas in every city or town with a population greater than twenty-five hundred, was introduced in the legislature in 1998, only to be defeated by private property interests. It is expected to be reintroduced later.[12]

Colorado's legislative efforts signal an aggressive attempt to put a halt to haphazard development stemming from the state's culture of uncompromising support for private property rights and concomitant lack of a robust planning tradition. Yet as bold as these initiatives appear, stateways rarely change folkways. Moreover, a top-down, centralized approach to land conservation has often failed to engage the local people whose lives and livelihoods ultimately depend on the land. Historically, states and the federal government have acquired and managed endangered natural resources, creating parks, refuges, and wilderness areas. This has tended to fragment conservation efforts among a variety of state and federal agencies, each with different management objectives and jurisdiction. As in the case of environmental laws, this resource-by-resource approach to land management has not accounted for ecosystems in their entirety, leading to unintended consequences such as rural sprawl.

Private landowners and local governments own and manage 80 percent of the nation's land. Half of the 728 listed endangered or threatened species under the Endangered Species Act are found on nonpublic lands. Seventy-three percent of the nation's forests (358 million acres) are under private ownership.[13] Private interests thus comprise the lion's share of control over the nation's natural resources. Accordingly, their decisions can often determine the quality of our environment.

In the light of this reality, many Coloradoans are coming to recognize that they themselves, and not the state legislature or federal government, must lead the effort to reverse the trend toward unplanned growth by creating communities that can provide for stewardship of the special places where they live and work and engender the values of neighborliness and civic responsibility that make that stewardship possible. They have come to understand that environmental protection efforts must take a system-wide approach, integrating many different interests and cutting across a variety of jurisdictional and physical boundaries. "When [the West] fully learns that cooperation, not rugged individualism, is the quality that most characterizes and preserves it," Wallace Stegner counseled, "then it will have achieved itself and outlived its origins. Then it has a chance to create a society to match its scenery."[14]

Heeding Stegner's advice, a diverse group of citizens, public officials, ranchers, and environmentalists has recently come together in Routt County to protect their land and way of life, charting a course toward livable, sustainable rural communities and a society equal to its scenery.

Cattle, Conservation, and the Fight for the Soul of Rural Colorado

Ranchers first came to Routt County in the later decades of the nineteenth century as part of the slow but steady settlement of the riparian valleys and uplands of the Colorado Plateau. Known for its beautiful viewscapes and rich, irrigated bottomlands, considered as productive as any hay lands in the state, Routt County retains much of the character of its historic rural landscape: a delicate tapestry of hayfields, trout streams, and cottonwoods spread out beneath the broad spruce- and granite-speckled peaks of northwestern Colorado. The Routt National Forest and the Mount Zirkel Wilderness Area provide county residents and visitors access to tens of thousands of acres of rugged backcountry. In the heart of the county is the ski resort town of Steamboat Springs. Located along the Yampa River and Interstate 40, Steamboat attracts over a million skiers a year and boasts a population of eight thousand people, accounting for roughly half the county's residents. With Olympic downhiller Billy Kidd as its chief promoter, the Steamboat Ski and Resort Corporation has become a major regional economic force over the past twenty years.

Driven by Steamboat's success as a world-class resort destination, intense development pressure has recently descended on the Yampa and nearby Elk River valleys as vacation homes and related development have begun to spread. The area's seasonal population has already increased threefold in the summer.[15] Between 1993 and 1994, county building permits increased 18 percent, while home values increased 36 percent. Today the average price of a single-family home in the Steamboat area hovers around $300,000, with developers paying up to twenty times the agricultural value of land.[16] Millions of baby boomers are expected to buy retirement homes in the coming years. Consequently Colorado's mountain resort areas like Steamboat are forecast to grow even faster than the state as a whole. Routt County, it is estimated, will grow by 66 percent in 2020, to 29,588 from 17,869 today.[17]

Concerned about the environmental and social changes that rapid growth has spawned in other parts of the region and throughout the Rocky Mountain states, local ranchers and conservationists, along with area residents, business owners, and government officials, have formed a unique alliance to develop and implement a community preservation strategy. Beginning in the early 1990s, they joined forces through a series of community planning efforts and private initiatives aimed at protecting the land and their rural way of life from the homogenizing, often corrosive forces of subdivision development and sprawl. With bumper stickers exclaiming clever soundbites like "Cows, not condos" and "Herefords, not highways," local ranching families, including some, like the Sherrods from the Elk River Valley, going on five generations, and their neighbors—elected officials, businesspeople, and conservation professionals—got together to figure out how they could use land protection as a way of not only ensuring the productivity and openness of the landscape, but preserving the values of community and responsibility that have defined the rural character of the Yampa and Elk River valleys for generations. They want to discredit the all-too-common refrain in the Rocky Mountain West that a rancher's last crop is a subdivision, and in the process seek to avoid becoming another Vail, the sprawling ski resort town a few hours' drive south in Eagle County.

As is often the case, the impetus for action came as a result of crisis. In the 1980s, a proposal to build a massive, Olympic-caliber ski resort called Catamount (named after nearby Lake Catamount) near the rugged Rabbit Ears Pass a few miles south of Steamboat ignited a firestorm between those who feared the opening of the county's floodgates to subdivisions, traffic congestion, and air pollution and those who welcomed more ski resort revenues for the county coffers. According Susan Dorsey Otis, one of the more vocal opponents of the Catamount proposal, who later went on to help establish the Yampa Valley Land Trust, a critical mass of civic consciousness was forged by the fight: "In the face of the Catamount struggle and other growth pressure in the county during the 1980s, with corrupt county commissioners offering up huge tracts to wealthy developers, we realized as citizens that we had to become proactive in our efforts to preserve the area."[18]

Arianthe Stettner, a city council member who fought the Catamount proposal alongside Otis and hundreds of other local residents, explains that as a result of the Catamount struggle, "we got educated. We eventually learned about public process through the Catamount debate. . . . We now know we have to be constructive, offer sound alternatives and be non-confrontational."[19]

The defeat of the Catamount proposal in 1991 set in motion a series of surveys and reports involving Routt County and Steamboat residents, businesses, and visitors to identify challenges posed by growth and development and the qualities that made the area a special place to live, work, and visit. For example, a survey conducted by the Colorado State University (CSU) Agricultural Extension Service in 1993 found that over 70 percent of Routt County registered voters preferred to protect 75 percent or more of the open ranch lands around Steamboat Springs, believing the open space and grazing cattle of the area's ranches contributed substantially to the county's unique identity and a sense of place.[20] A 1993 report on visitors' attitudes toward ranch lands in the county found that for nearly eight out of ten visitors, open space significantly increased their summer visits and willingness to pay for the quality of the experience. When asked if the conversion of ranch lands to other uses would influence their decision to visit the area, 50 percent said they would seek other resort destinations.[21] Similarly, a 1994 report by the Vision 2020 project, a multiyear planning and visioning effort spearheaded by a diverse group of Routt County citizens, singled out the preservation of open space as the community's top priority and among the most important factors in determining the area's overall quality of life.[22]

Meanwhile, in an effort to establish stronger ties between ranchers and environmentalists in the county, two groups with a unique stake in open space protection, Susan Dorsey Otis and C. J. Mucklow, CSU's Routt County agricultural extension agent, organized an "Eco-Ag" committee in 1995. Traditionally environmentalists and ranchers have considered each other archenemies. Environmentalists have consistently opposed ranching activities, claiming that cattle harm fragile western ecosystems by overgrazing vegetation and soils and destroying stream banks. As well, they have promoted a host of federal environmental policies considered anathema to ranchers, from restrictions on grazing permits in national forests to the reintroduction of wolves into wilderness areas in Wyoming and Montana. Some environmentalists have even resorted to cutting the barbed wire fences on ranchers' property in protest against allowing cattle to roam on federal lands and the resultant environmental damage.[23]

For their part, ranchers have resented environmentalists' resistance to their way of life and viewed restrictions on their grazing permits as subversive, an example of unnecessary government regulation and infringement on their private property rights. Further, the reintroduction of wolves has meant the possibility of livestock depredation, which for ranchers is a direct assault on their property and way of life. Environmentalists, in ranchers' eyes, are elitist do-gooders, forever meddling in other people's affairs.

Though at first the Routt County ranchers and environmentalists were wary of each other, over time they were able to find common ground through informal meetings and lots of cups of coffee in ranchers' kitchens. Lynne Sherrod, a local rancher who attended these meetings, attributes the

Eco-Ag committee's success to the willingness of the parties to see themselves as neighbors and "move outside [their] own comfort zone" in investing trust in one another. "The reason people take risks," Sherrod explains, "is because they're invested and accountable. Conservation is all about who the people are."[24]

Among the environmentalists whom Otis and Mucklow brought to the table were representatives from the Nature Conservancy (TNC), a national nonprofit land conservation group headquartered in Washington, D.C., which had opened an office in Steamboat in 1992. TNC program manager Jamie Williams remembers that initially ranchers perceived of his environmental group as as much of a threat as developers. "The big fear was that we would come in, buy up the valley, take land out of production and erode the local agricultural community," he says, referring to TNC's traditional core strategy of purchasing large tracts of land and then prohibiting most human activities on them, much to the dismay of ranchers.[25] In the ten western states, the group has already protected 3.4 million acres of private land, mostly through direct acquisition.

Realizing the limits inherent in its acquisition-based approach to land conservation, in terms of both financial resources and land management, TNC has begun to look to partnerships and working landscapes as an alternative. As Williams explains, "We can't really achieve our mission through land acquisition, and it wouldn't be desirable anyway. It's really driven the Conservancy to expand from a purely land-acquisition focus to one of building partnerships with private landowners and other people. . . . The common ground [is] the tidal wave of development that [is] threatening. . . . Our ultimate goal . . . is to support other stewards of the land on the land."[26]

Mark Burget, TNC's Colorado state director, suggests that TNC has arrived at a new vision for its work, one that is right at home on the ranch and working landscapes generally. "We have to realize that while we have a lot to offer in the way of education about ecosystems and ecosystem management, we have a lot to learn about how best to leverage our role and build community in the process."[27] Burget explains that TNC's new philosophy is explicitly geared toward community and economy, not just the environment, and that these three concepts taken together are the "three-legged stool" representing the required elements for stability and strength in any sustainable structure.

According to Greg Low, director of TNC's Center for Compatible Economic Development, the commitment to people and livelihoods as part of the organization's conservation mission came from the realization that "conservation goals could not be securely attained otherwise. We realized that long-term conservation of significant natural environments will succeed only with strong support from the people who live and work in these areas. . . . We learned that examples of development that protect the environment, generate profits and enhance quality of life are rare. Therefore, we decided that we must work to foster successful, locally-based compatible development initiatives."[28]

Burget points to TNC's burgeoning partnerships with Yampa and Elk River Valley ranchers as an example of this new vision. Where once TNC staff would not be allowed anywhere near ranchers' property, today Yampa River program manager Mike Tetreault and TNC scientists are treated as good neighbors, having established a lasting bond of trust. Now they can work with ranchers on such conservation projects as nourishing stream banks, preventing soil erosion, and regenerating cottonwoods, willows, and other riparian vegetation essential to the health of the area's ecosystems.

Elk River Valley rancher Jay Fetcher, a second-generation cattlemen whose father, ironically, helped start the Steamboat ski area at about the same time he purchased the family's thirteen-hundred-acre ranch in the 1950s, recalls that his conversion to conservationism came about as a result of his "belief that [his] ranch is a productive ranch worth saving, just as food production is a valuable part of the local economy and culture."[29] Fetcher and five other ranch families from the area, including Lynne Sherrod's, came together on their own in 1992 to devise a strategy for protecting open space and the county's rural character, recognizing that environmental protection was the key to preserving the area's agricultural economy.

Blending their agricultural, economic, and cultural concerns with a conservation and open space focus, the ranchers formulated a conservation-based development strategy to protect the area's rural heritage and ecology. They wanted to protect in perpetuity the open and productive character of the area that comprises the basis of its economic vitality. Unlike traditional conservation efforts, they were intent on protecting the area as a whole, not just islands of land, with working landscapes as a main feature. They saw conservation and development not as opposites or mutually exclusive, but as flip sides of the same coin.

Marty Zeller, a landscape planner who in the 1970s and 1980s got involved in innovative land protection programs in Vermont and Massachusetts, provided valuable professional support to the ranchers from the start. His group, Denver-based Conservation Partners, assisted the ranchers in designing their conservation-based development strategy, whose primary objective, he explains, "is the protection of the agricultural land in the valley so that any new development must minimize its impact on these lands and the agricultural activities necessary to use them productively."[30] Other conservation objectives identified include protection of the valley's open, dramatic scenery and wildlife. As Zeller states, the strategy is intended to preclude large-lot development, which consumes large tracts of productive agricultural land and disrupts wildlife habitat. Moreover, Zeller points out, the strategy is meant to replace traditional zoning measures because zoning is an impermanent land protection technique, subject to changing political and economic conditions.

The development side of the ranchers' strategy consists of two components: reserved and limited development and affordable housing settlement. *Reserved development* denotes the development of sites that are located in areas that do not interfere with wildlife habitat or scenic or agricultural values. *Limited development* refers to the sensitive development of a portion of property in exchange for the protection of the remainder through the use of a conservation easement, a deed restriction permanently limiting the landowner's right to subdivide property into smaller parcels while maintaining private ownership and economic use of the land. Because a conservation easement reduces the development potential of land, thus stripping it of some, if not most, of its market value, the landowner, while denied a greater selling price, enjoys the benefit of reduced estate taxes. However, under a limited development approach, a conservation easement allows the landowner to achieve considerable economic gains through the creation and sale of homesites in strategic locations, such as vegetated, south-facing slopes that provide ample sunlight but minimize visibility, or set back from the most productive agricultural lands.

In order to ensure that all development fits the landscape to the greatest extent practicable, other measures can be taken in addition to site location.

Design guidelines for limited development encourage the use of dark, natural colors and building materials that blend with the surroundings, prohibit reflective surfaces, limit the size and scale of structures, and define appropriate architectural styles.

The affordable housing component of the development plan entails the creation of a small number of affordable housing sites in a clustered, village location for those who work in the area but have limited access to the housing market. Affordable housing is required because the ranchers' conservation measures will necessarily limit the availability of affordable housing sites, while likely increasing property values even beyond already high rates. Affordable housing sites will be targeted to developed areas, such as the unincorporated village of Clark, with 455 residents, in the Elk River Valley, to take advantage of access to the village general store and a community meeting place.

Zeller and the Elk River Valley ranchers set down their conservation and development strategy in the Upper Elk River Valley Compact, a nonbinding statement of principles agreed to by ranchers like Fetcher, Sherrod, and their neighbors. The one-page document spells out the ranchers' motives and goals in adopting the compact:

> We believe that the valley is a special place in terms of people, landscape and culture. As landowners, we share a commitment to protect the special scenic, rural and working character of this valley as well as to enhance its agricultural viability. We wish to pass on to future generations a valley that possesses the special qualities we enjoy today.
>
> In order to do this, we believe cooperative action between landowners is essential. As responsible members of this community, we prefer to create a future that responds to our shared objectives rather than be bystanders in a process that will inevitably lead to an erosion of the special character of the valley. Though our conservation motivations and economic circumstances are as varied as the family and personal reasons that brought us to the valley, we share a common set of principles that should shape future actions and decisions within the valley.[31]

The compact provides a set of planning and implementation principles that establish the general framework for achieving the ranchers' goals. The planning principles stress protecting the valley's rural character by supporting productive rural land uses, specifically ranching, agriculture, and low-impact recreation, guiding new development away from prime agricultural and forest lands and providing compact, affordable housing settlement. These principles, the document states, "should create a special image and competitive advantage for the valley."

The implementation principles encourage techniques that can be flexibly tailored to meet the economic and conservation needs of each landowner, such as conservation easements donated to qualified land conservation or land trust organizations like TNC or the Yampa Valley Land Trust. The principles emphasize the power of voluntary and cooperative action, as well as governmental policy that provides incentives to such action. In the event the landowners have to dispose of their properties, the principles call on them to seek buyers who share their conservation and agricultural objectives. In addition, the principles urge "effective interaction with the local community, the county and state and federal land management agencies to ensure that the land protection and agricultural objectives of the valley are taken into account when decisions are made that affect the future of the valley," and invite the support and cooperation of neighbors and elected officials in achieving the compact's goals.

Rancher Jay Fetcher donated 1,250 acres to a local land trust and helped establish the Colorado Cattlemen's Land Trust (CCLT), the first land trust designed specifically by and for ranchers and headed by Fetcher's neighbor, rancher Lynne Sherrod. The CCLT currently protects over 22,000 acres in Colorado. Another Elk River Valley neighbor, the Stranahans, placed a 700-acre easement on their property. Down the road in the Yampa Valley, rancher Dean Rossi part-sold, part-donated a conservation easement to the Yampa Valley Land Trust, protecting four miles of the Yampa River and its unique riparian forest. TNC will help manage the habitat.

Zeller's and the ranchers' innovative work in the Elk River Valley dovetailed perfectly with the efforts of the larger, county-wide group established during the Catamount fight and aimed at establishing a comprehensive land use plan. Buffeted by the momentum of these parallel projects, the Routt County Open Lands Steering Committee was formed in late 1994 under the auspices of the Routt County Board of County Commissioners. Its mission was to develop a list of strategies for protecting the county's open lands—its productive agricultural, natural, and scenic lands—and to engage the community, especially landowners, in the process. The steering committee was chaired by Susan Dorsey Otis and composed of a diverse group of citizens: ranchers, bankers, contractors, environmentalists, land planners, and government officials. The county planner and attorney, extension agent C. J. Mucklow, the Steamboat city planner, and Zeller provided technical support to the steering committee.

After extensive community meetings over nine months, the steering committee issued the Routt County Open Lands Plan in early 1995, which, like the Elk River Valley Compact, was authored by Marty Zeller. Funded by proceeds from the state lottery administered by Great Outdoors Colorado (GOCO), the state environmental grant-making agency, the plan takes to scale the principles set forth in the compact and proposes a set of eight recommendations for protecting the county's environment, culture, and economy that rely on voluntary, incentive-based approaches rather than command-and-control regulations. The plan recommends:

1. Adopting a statement on the importance of agricultural lands, natural areas, and open space to Routt County;
2. Enacting a Routt County Right-to-Farm and Ranch ordinance, providing protection to agricultural operations from nuisance complaints (for example, odors);
3. Encouraging a private technical resource team to help landowners select and tailor the land protection options available to them;
4. Establishing a voluntary conservation easement-with-homesites program to enable landowners to profit from real estate sales while protecting portions of the property;
5. Establishing an Open Land Subdivision, which encourages landowners to donate conservation easements through a subdivision provision in exchange for being allowed to build on smaller lots;
6. Establishing a Land Protection Subdivision, to encourage both cluster development and development which is sensitively designed for the landscape in exchange for a small increase in density;
7. Establishing a Purchase of Development Rights (PDR) program using public funds for the voluntary acquisition of development rights from key ranch lands;
8. Establishing a Transfer of Development Rights program to trade development rights in sensitive areas for increased development in designated growth areas.

The county commission adopted the recommendations, which became part of the county's master plan in June, 1995. To fund the PDR program, Routt County voters passed a small property tax increase in 1996, amounting to roughly $360,000 a year. GOCO contributed an additional $250,000 to purchase the first round of easements, which both Steamboat and the county matched. Further, the Yampa River System Legacy Project, a companion project to the Routt County Open Lands Plan, received a grant of

$6 million in 1996 from the lottery-funded GOCO. The Legacy Project was created to "protect and enhance the ecological health of the Yampa River and the productive agricultural lands it supports while providing for appropriate recreational opportunities."[32] The Legacy Project expanded the county's already ambitious goals while increasing its capacity to achieve them.

Other initiatives that spun off the Open Lands Plan process included a community comprehensive plan, an intermodal transportation plan, an affordable housing plan, and an economic development plan. Given all the brainstorming and dialogue that gave rise to these plans, many county residents joked that people were "blinded by vision" and enervated by the seemingly endless visioning sessions.[33]

The Open Lands Plan is the public policy counterpart to the ranchers' private conservation-based development strategy. It institutionalizes and enhances that strategy, incorporating publicly financed techniques like purchase of development rights (PDR) and providing incentives for voluntary open space protection. The plan has met with extraordinary success to date, ushering in what many are confident will be a new era of careful, sustainable growth and development in Routt County. "We've accomplished a lot using the plan," explains Routt County planner Andy Baur, "It's been a critical document because it shows that the community and its stakeholders are serious about preserving open space. It's a mandate for people to take planning seriously, and it gives us something truly useful to show developers, ranchers and other property owners who need different scenarios." Marty Zeller echoes those sentiments: "The plan struck a chord in Routt County, where development pressures are enormous. The timing was right. But what made the plan work was the real leadership behind it. . . . Nowhere else have I seen all these forces come together like this."[34]

Yet in the light of increasing land values and perennially unstable beef prices, the temptation for ranchers to sell their land is still strong. Even with the Open Lands Plan, the Legacy Project, and the subsidies provided by GOCO, the county, and Steamboat, conservation easements cannot compete with developers dollar for dollar. And not every rancher, given the choice, would choose to do what the Fetchers, Stranahans, and Rossis did.

In an effort to try to level the playing field, TNC, C. J. Mucklow, and their colleagues are developing value-added agricultural enterprises to help keep land in agriculture. For example, in late 1998 they rolled out the

Yampa Valley Beef program. Market research conducted for TNC's Center for Compatible Economic Development found that most consumers in Routt County are willing to pay a premium for beef raised in an environmentally responsible manner. That premium, 20 percent above the cost of generic commodity beef, would go to participating ranchers, providing a financial incentive for cattle ranching in the face of competition from developers seeking to subdivide ranch lands. In one year, a rancher with twenty-five cattle can expect to earn an extra $1,500 from participating in the program.[35]

"Our niche is the land-conservation angle and trying to have a wholesome, natural product," says Mucklow, the beef program coordinator.[36] Ranchers began selling the beef in December 1998 through local restaurants and markets, hoping to create a new brand choice, on par with other environmentally responsible brands like Ben & Jerry's or Tom's of Maine. TNC has also urged that at least 25 percent of the cattle in the program be raised or grazed on land that is protected through a conservation easement, and some of the program revenue will be donated to local land trusts.

Meanwhile, Mucklow has helped to develop a Routt County lamb's wool blanket business as a way of adding value to ranching operations. Sales of the product have gone from a few dozen to several hundred within two years. These and other enterprises are market-based strategies meant to complement the Open Lands Plan and to counter powerful market pressures for ranch lands development. They are part of the tool box that Routt County residents have at their disposal to help them achieve their conservation-based development goals.

Even more important than the variety of innovative strategies Routt County residents have devised is the civic will that undergirds them. All of the diverse stakeholders in Routt County were and remain committed to being good neighbors and good citizens, building a genuine community and protecting the physical setting, the place, and living by certain values known and agreed to by that community. To affirm their civic commitments and minimize the potential conflict between new and existing residents, C. J. Mucklow and over a dozen Routt County citizens published *A Guide to Rural Living and Small-Scale Agriculture,* a concise booklet outlining the basic facts and requirements of life in Routt County, from building codes and open burning management, to outdoor etiquette and safety, to hay production and growing seasons. The guide does not seek to be proscriptive. Rather, it articulates the ways in which Routt County is a special place, a place worth living in and caring about, and what people can do to keep it that way. As the saying goes, "You can't have rural character without rural characters." Routt County residents take that maxim seriously.

Civic Environmentalism and Routt County's Conservation-Based Development Efforts

Routt County residents' success in creating and implementing their comprehensive conservation and development strategy stems from their civic environmental approach. In the face of the proposed Catamount ski resort and other growth pressures, ranchers, environmentalists, city and county officials, businesspeople, and others were willing to engage in a comprehensive community planning and visioning process and to seek appropriate professional support to help them devise innovative strategies. As Vision 2020 chair Arianthe Stettner described, the process itself became an opportunity for area residents to learn about how to take control of their shared destiny as a community and place.

Through surveys, reports, and meetings, the diverse group of stakeholders identified the social, economic, and environmental assets that make Routt County a unique, desirable place to live, work, and engage in recreation. The group pinpointed environmental quality as the key to the county's overall quality of life and designed strategies linking the conservation of open lands and ecosystems to productive landscapes, agriculture, recreation, and tourism. In effect, by mixing private and public approaches to conservation and development, evidenced by the ranchers' conservation easements on the one hand and the Open Lands Plan on the other, Steamboat and Routt County as a whole have had to look beyond narrow legal or regulatory responses to their environmental and social challenges and have engaged an assortment of flexible, responsive techniques. Environmental laws were not designed to address the kinds of land use and development problems county residents face and so have little to do with their overall approach.

Further, TNC, in redesigning its mission to serve Routt County residents and the local economy better, has been able to work effectively with ranchers to educate them about ecosystem management, capitalizing on their shared concern for productive ranch lands and riparian forests. TNC, for its part, has come to rely on ranchers for environmental leadership, creating a win-win situation.

Moreover, in their concern for preserving jobs and the local economy, ensuring affordable housing for workers, and reaching out to and including a diverse array of community residents, the people at the center of the community preservation movement, among them Susan Dorsey Otis, C. J. Mucklow, Jamie Williams, Jay Fetcher, and Lynne Sherrod, have demonstrated a commitment to the civic democratic principles of environmental justice. Although there are few people of color in Routt County, save for the migrant ranch hands and ski resort workers who pass in and out of the area, there is poverty and disaffection. The Routt County citizens showed a sensitivity to these issues and designed their strategy accordingly.

Regarding the ecoindustrial dimension of civic environmentalism, the Yampa Beef Project and TNC's embrace of ecologically minded agriculture represent an ecoindustrial approach to ranching. With few toxic inputs such as pesticides or other hazardous materials, ranching's greatest environmental impacts stem from the cattle themselves, whose grazing can harm stream banks and woodlands. A truly ecoindustrial system would involve the use of renewable energy to power farm equipment and facilities, and the recycling of all ranch-generated wastes, among other measures; the ranchers' commitment to land preservation and environmentally responsible livestock management is a step in the right direction toward sustainability.

Perhaps the most compelling civic environmental aspect of the Routt County land protection effort is the concept of place. The cottonwood forests that line the stream banks, the delicate soils of the ranch lands still marked by the wagon wheels of nineteenth-century settlers, the granite peaks and verdant blanket of aspen and spruce, the vast, open spaces reaching up to the night sky: more than just physical wonders, these places comprise Routt County's culture and economy, infuse the consciousness of its residents, and inspire a commitment to something larger than themselves. This is Wallace Stegner's West, the West of the imagination. Place seems to have an emboldening effect on the people of Routt County; they are willing to experiment. In Routt County, the place and the community are inseparable and mutually constituting.

As Daniel Kemmis has written about another unique corner of the Rocky Mountains, his home state of Montana, "Public life can only be reclaimed by understanding, and then practicing, its connection to real, identifiable places. . . . The political culture of a place is not something apart from the place itself. By exactly the same token, the strengthening of political culture, the reclaiming of a vital and effective sense of what is public, must take place and must be studied in the context of very specific places and of the people who struggle to live well in such places."[37]

Routt County residents came together to plan and take action because of their devotion to their shared and beloved place, initiating what is hoped will be a perpetual cycle of community preservation, of people and place. They are living out what journalist Timothy Egan describes as a new western ethic: "the idea of letting th[e] land be itself."[38]

Yet notwithstanding their remarkable success to date, Routt County faces an uncertain future. In the light of the area's skyrocketing real estate values and the market forces that drive any ski resort economy, especially in Colorado, Routt County has to contend with a formidable challenge: maintaining an affordable cost of living for area residents. The county must also reckon with the paradox that attends its farsighted land conservation strategy. Just as the strategy will reduce the supply of land available for development, thus raising the value of existing developed and buildable land, so too does conservation itself increase the value of land. Access or proximity to conservation land is considered a real estate amenity for which most buyers are willing to pay a premium. These unintended consequences of land protection can be managed by leveraging the income and property tax revenues from large landowners and more affluent residents to ensure an adequate supply of affordable housing and provide high-quality municipal services and public education to all area residents.

To the extent that residents remain vigilant and engaged, these challenges are surmountable. They have established the personal relationships and shared sense of mission that are central to creating civic alchemy and lasting environmental solutions. Given their success so far, it is likely the people in Routt County will continue on the path to an economically stable, wide-open place well worth protecting.

7

Smart Growth, Community Planning, and Cooperation in Suburban New Jersey

[Open space] seems to foster unnatural alliances, making back-slapping buddies out of predators and prey.

—Angelo Morresi, in the *Newark Star-Ledger*

We have got to understand that once the land is gone, it's gone forever.

—Christine Todd Whitman, in *The New York Times*

The Land of the Suburb and Relentless Growth

New Jersey. To many, the very name conjures up images of rush-hour gridlock, highway exits to nowhere, and industrial facilities spewing lethal chemicals into the state's air, water, and soil. Routinely the butt of environmental jokes, New Jersey has long suffered the reputation as the nation's largest, most densely populated toxic waste site. The state's formidable industrial heritage, reflected in the many abandoned factories that languish in aging cities like Newark, Trenton, Jersey City, and Camden, and its proximity to the major northeastern metropolitan centers of New York and Philadelphia, have made the state about as polluted as it is populous (population approximately 7.75 million). As one former member of the New Jersey State Planning Commission remarked, New Jersey "is the nation's most densely populated state, and therefore, among the first to encounter crises of garbage, degradation of its water supplies and chemical pollution."[1]

It was not always this way. Nicknamed the Garden State, New Jersey's industrial past is matched by its rich agricultural history. The state's numerous farms supplied most of the fresh vegetables sold at markets in the New York metropolitan area and beyond. Produce was also grown off-season in commercial hothouses throughout the state.[2] Venture beyond the borders of the state's many major highway corridors, and there are plenty of vestiges of New Jersey's robust agrarian tradition.

Ironically, it was the state's farmland and pastoral beauty that brought the nation's first suburbs to New Jersey, the same suburbs that eventually would overwhelm the state's rural landscape and result in the sprawl and pollution associated with today's New Jersey. With the advent of the railroad in the mid-1800s, the practicality of commuting between cities and outlying areas in the Northeast was greatly enhanced, as the constraints of time and poor road conditions which attended travel by horse-drawn coach were all but extinguished. Suddenly businessmen who had no choice but to live in crowded, polluted cities like New York and Boston, where they worked, could move to the relatively pristine, sparsely populated countryside as long as a railroad station was nearby.

In 1853, Llewellyn Haskell, a New York City pharmaceutical merchant who lived along the Passaic River in New Jersey (commuting to work via coach and ferry boat), decided to move farther west of the city when he began suffering health problems and suspected that the bad air and salt marshes near his home were to blame. He settled on the village of West Orange, known for its wild scenery, expansive views, and clean air. Located a distant twelve miles from his business in New York, the new Morris & Essex railroad line, with a station stop at West Orange, made it possible for Haskell to commute. He purchased a 40-acre farm with distant views to Manhattan. A few years later, in 1858, Haskell bought an additional 350 acres with the intent of developing a residential community designed to make the most of the area's natural beauty. With rocky outcroppings, ravines, tree-shaded roads, and rustic pavilions, Llewellyn Park was established as the nation's first railroad suburb.[3] It was as exclusive as it was expensive, with lot sizes ranging from a minimum of 1 acre up to 20 acres. Eventually, Haskell, realizing the substantial profits to be made in suburban real estate ventures, acquired another 400 acres, and Llewellyn Park became known as one of the region's fanciest communities.

But as author James Kunstler explains, Llewellyn Park, like so many other modern suburbs, was not really a community at all. Although it offered a rustic and romantic setting, "it couldn't have been more artificial. It lacked almost everything that a real community needs to be organically whole: productive work, markets, cultural institutions, different classes of people. And the houses were so far apart that the residents would lose all awareness of their neighbors."[4]

Roughly a hundred years after Llewellyn Park was established, New Jersey and the rest of the Northeast experienced unprecedented suburbanization following World War II. Federal mortgage subsidies, the GI Bill, and major highway projects encouraged the creation of a market in new housing construction in formerly undeveloped areas. Lewis Mumford, the eminent urban historian and one of the founders of the Regional Planning Association of America, characterized this new growth as the fourth great migration of American history: a decentralized, sprawling movement of people and infrastructure away from cities. In Mumford's analysis, the first migration was the settlement of the frontier in the nineteenth century; the second, the development of towns and cities; and the third, the migration of Americans from farms to urban centers.[5] Echoing James Kunstler's criticism of Llewellyn Park, Mumford worried about the consequences of new, large-scale suburban development, explaining that the "pedestrian scale of the suburb disappeared, and with it, most of its individuality and charm. The suburb ceased to be a neighborhood unit; it became a low density mass."[6]

Despite Mumford's concerns, suburban development burgeoned in the postwar United States, and particularly in the populous Northeast. Across the New York Harbor from New Jersey, in Levittown, Long Island, the first truly modern suburb was developed in 1947 on what was formerly farmland, consisting of 17,000 single-family homes packed together in neat rows serviced by countless miles of winding streets. With the Levittown model in place, farm after farm in New Jersey and throughout the rest of the Northeast was sold to developers and converted to subdivisions virtually overnight. With the subdivisions came shopping centers, gas stations, and schools and, in their wake, streams of traffic and other environmental assaults.

In the three decades between 1950 and 1980, New Jersey was transformed from a largely rural landscape of farms, villages, and towns into a massive suburb, from the Garden State into a place best known for its eponymously named turnpike. Because of the importance of highway construction to the growth of suburbs, the New Jersey Turnpike is perhaps the best symbol of the state's suburbanization and environmental change. With up to fourteen lanes in some sections and recent lane expansion costing over $3 billion, the turnpike is a physical and financial behemoth. Lacking a comprehensive transportation plan or urban policy, the state of New Jersey and the federal government have spent billions of dollars over the decades building thousands of miles of highways across the state's countryside, simultaneously drawing people away from urban centers and toward new subdivisions.

New Jersey's highways alone account for 45 percent of the nation's total traffic tolls.[7] Each year, the millions of cars and trucks traveling those highways emit 1.5 billion tons of carbon monoxide into the air, just one of several transportation-related pollutants. That figure is expected to reach 2.9 billion tons in 2010.[8] Because of the state's poor air quality, it no longer qualifies for federal highway funds, thus necessitating the huge toll revenues. The result is a degenerative cycle of roadway construction, air pollution, and sprawl.

With real estate speculators eager to buy up all buildable land, unprotected hillsides, fields, wetlands, and streams have become prime sites for residential and commercial development. Close to two-thirds of the state's 4.8 million acres have been developed; only about 30,000 acres are protected as critical habitat. Since 1950, New Jersey has lost more than half its farmland; as of 1990, only 870,000 acres remained. According to a 1992 Rutgers University study of land use patterns in New Jersey, if current trends continue, the state will lose another 108,000 acres of agricultural land and, in the process, generate 15,160 tons of water pollution (nitrogen, phosphorus, lead, zinc), from contaminated runoff due to suburbanization.[9]

The intensity and rate of growth in New Jersey, and the widespread perception that the state's environment and quality of life are on the road to ruin, have inspired a lot of soul searching among New Jersey residents. All the bad jokes seem to have finally a struck a nerve. For example, in November 1998, New Jersey voters passed an ambitious $1 billion plan to save vanishing farmland from development, including a constitutional amendment that dedicates up to $98 million annually for open space conservation.

Why, one might wonder, has it taken so long? Suburbanization itself seems to be part of the reason. First, suburban residents have tended to be content about their lot in life. Complacency has been the norm. Moreover, as suburban critics like Mumford and Kunstler suggest, the physical layout of suburbs—their very structure and design—tends not to foster a sense of community, place, and common fate that is central to preventive social action. In other words, suburbs seem to have an insulating and inhibiting effect on individuals' civic will. By being segregated on single lots and forced to depend on automobiles for travel, suburban residents are largely deprived of the opportunity to act collectively and responsively in the face of adverse environmental conditions such as air pollution, traffic congestion, and loss of open space, to which they themselves principally contribute.

And there lies the paradox of suburban life: the desire for privacy, which often motivates people to move to or stay in suburbs, ultimately undermines their competing desire for environmental quality. This is a variation on the ecologist Garret Hardin's notion of the "tragedy of the commons," in which individual actors, pursuing individual ends, destroy their common natural resources through overconsumption, hurting everyone in the end.[10] For Hardin, the key to environmental protection is the capacity of individuals to work together toward common goals.

In spite of the stasis engendered by suburbanization, many New Jerseyans have mustered remarkable zeal and creativity over the past decade in working to establish a strategy for creating a sustainable state—a new model for achieving environmental quality, economic prosperity, and social health. Across New Jersey, examples of civic environmental action are taking shape, making the state a forerunner in the growing movement toward sustainability.

Protecting Deer, Ball Fields, and Open Space in Morris County

Located in Morris County in the northern part of the state, Randolph Township is a quintessential suburban village about forty miles due west from Manhattan and twenty-five miles northwest of Haskell's Llewellyn Park in West Orange, an easy ride away on Interstate 280. A middle-class Manhattan commuter town, Randolph has a population of 24,000 and, according to area residents, is bursting at the seams. Ball fields are crowded seven days a week, schools are packed with more children than local property taxes will support, white-tailed deer, lacking sufficient contiguous woodland habitat, prance from lawn to lawn in search of tasty flower beds, and once-quiet residential streets are often clogged with traffic.

Beginning in the early 1990s, Randolph residents passed a referendum levying a two-cent property tax to help fund the purchase of open space, known as the Open Space Trust Fund. These funds are part of a larger open space finance program that consists of Morris County's open space fund, financed by a separate two-cent county tax, and state loans and grants, financed by land preservation bonds, known as the Green Acres program. Established in 1961, the Green Acres program is the precursor to the municipal and county efforts that have blossomed in the past ten years across the state. With thirteen counties and fifty-three towns adopting open space initiatives starting in 1989, New Jersey is the country's leader in the number of jurisdictions where voters have approved open space taxes.[11] Unfortunately, the situation has become so critical that New Jersey voters appear to have no choice.

The average Randolph home owner, with a real estate tax assessment of $128,000, pays $25.60 a year in open space taxes, on top of the regular town, county, and school property taxes. Resident Anita Calotta, a champion of open space protection, says she is more than willing to pay the additional tax to ensure that her family and neighbors have access to parks and woodlands. "The more land the town can gobble up, the better," she exclaims, "I'll pay." Another Randolph resident, Melinda Kryger, is as sympathetic to the deer who invade her front yard as she is supportive of the open space tax: "They ate my azaleas, my tulips and my rhododendrons. It's like a salad bar to them. We keep invading their space and they have no place to go. I'll pay to keep the land wooded."[12]

To date, Randolph Township has been able to generate over $5 million in open space funding, which, according to assistant town manager Thomas Czerniecki, has been used to purchase over 300 acres of woods and parks. Coupled with existing green space, this gives the town roughly 900 protected acres.[13] Randolph's open space now comprises roughly 11 percent of its total land area of 13,400 acres, from 8.6 percent before the purchases. The biggest piece of land acquired so far is 172 acres of woodland abutting a residential subdivision. The property links playgrounds and ball fields to 600 acres of county-owned woods, resulting in a significant greenbelt for the town.

As important as funding is for open space protection, equally important is the process by which the town decides what land to purchase and how to manage it. In 1994, local residents pushed for the creation of the Randolph Township Open Space Committee to ensure citizen participation and strategic decision making. Comprising 13 citizen-volunteers, the committee developed a ranking system to determine what sites make the most sense to protect. The criteria include environmental or habitat value, development pressure, size, and connectivity or proximity to other open space. Barbara Davis, chair of the committee, says the open space program enjoys widespread support because the local community itself drives the decision making. "The citizens are in charge, and work hand-in-hand with town officials," she states proudly. "We know what we want our town to look like, to feel like, and we are committed to getting results that benefit the whole community." She adds, "A lot of people felt that Randolph was becoming a community they no longer wanted to live in, given the skyrocketing real estate prices and land grab. Our committee, made up of farmers, lawyers, bankers, and other smart, dedicated residents, is able to hone in on what the community wants. Open space is only as good as it is defined by real people in a real community."[14]

Assistant town manager Czerniecki adds that adequate staff resources, such as a full-time town planner, are essential to managing publicly owned land. It is not enough, he explains, simply to purchase property; the town and its citizens must commit to looking after it. Czerniecki also points out that a vital companion piece to the preservation of open space is the provision of affordable housing for the community's lower-income residents. Because proximity to open space and parks increases property values while reducing the amount of developable land, towns must be vigilant in ensuring that affordable housing is provided as a component of any open space protection strategy. As a result of a 1985 court case involving affordable housing in the town of Mt. Laurel, state law requires that municipalities provide an adequate supply of housing for lower-income residents. However, many towns get around the requirement by bargaining away up to half of their affordable housing obligations to neighboring jurisdictions in favor of more open space protection and higher property values.

Moreover, developers of affordable housing often seek to build mixed housing subdivisions, containing both market and subsidized units, on large tracts of open land. This practice is known as the builders' remedy because it satisfies the Mt. Laurel affordable housing requirement while generating greater profits through the sale of market-priced units. However, the sheer size of these subdivisions, containing both affordable and market units, results in significant loss of open space, thus pitting open space protection against affordable housing development.

Randolph residents have chosen to keep their community open to lower-income residents and, claims assistant town manager Czerniecki, have worked hard to fulfill their obligations under state law to ensure that a portion of the housing market remains affordable while still protecting open space. Nevertheless, he warns, maintaining affordability for the community at large is an ongoing challenge.

The town of Randolph's open space efforts, and those of other Morris County municipalities, have been aided by Morris 2000, a nonprofit organization founded in 1984 to identify and pursue quality-of-life goals for the county for the year 2000. The organization promotes multi-community efforts aimed at achieving sustainable economic growth, livability and a sense of community, and environmental protection throughout the county. It provides a forum for county residents, elected officials, civic organizations, and businesses to build consensus and foster leadership, and it educates stakeholders on specific social and environmental policy issues. In 1984, 1990, and 1998, the group hosted visioning and strategic planning sessions at which the question was posed, "What do you want Morris County to look like in the years ahead?" Soliciting responses to this question, and then designing strategies around the responses, is what Morris 2000 is all about.

Morris 2000 was responsible for placing the county's first open space tax on the ballot in 1992, and it has actively supported the establishment of open space initiatives in Morris County municipalities like Randolph Township. Further, working with twenty-four of the county's thirty-nine municipalities, the group has helped develop two major watershed protection initiatives, the Ten Towns Great Swamp Watershed Commission and the Rockaway River Watershed Cabinet, whose mission is to engage citizens in water quality monitoring and promote environmentally sensitive development. For example, Morris 2000 has worked with the Watershed Cabinet to encourage the redevelopment of several brownfield sites along the Rockaway River as a way of leveraging the open space efforts underway throughout the county and restoring the historically industrial river (it was the site of several iron mines dating back to the mid-1800s) to its rightful place as an important environmental and economic resource.

Similarly, Morris 2000 has been instrumental in helping to push for transit-based development in several town centers located along the new MidTown Direct commuter rail, which for the first time shuttles passengers nonstop from Morris County towns to New York's Penn Station. Increasingly commuters are looking to buy houses near the commuter rail stations in towns such as Dover and Morristown, thus reducing the need to build more housing on farmland and open space and expand the capacity of county roadways. As well, through the Housing Partnership for Morris County, a project Morris 2000 started in 1990, the group is investigating ways to increase affordable housing opportunities in the county while avoiding the open space threats associated with the builders' remedy.

Morris 2000's executive director, Carol Rufener, explains that the key to achieving environmentally and economically sustainable communities in the county is to engage as many different stakeholders as possible in the process of determining the county's future—where it should be headed and how to get there. "We plan locally, but live regionally," she explains, referring to the need for county-wide, system-wide planning and action. "We must create regional solutions to our environmental and social problems that engage our citizens."[15] Rufener's group sees itself as a "catalyst for implementation" and "instrument for consensus building," both essential to seizing control of the county's fate and ensuring responsible, or smart, growth.

Randolph Township and Morris County's success at protecting open space and managing growth has been replicated across the state, as more and more towns and counties, totaling fifty-three and thirteen respectively, have adopted open space programs. Now mayors routinely tout the benefits of penny taxes for open space acquisition as farsighted fiscal alternatives to the significant property tax increases that would be required to support new schools and roads and more police and other municipal services. The buy-in and support of local citizens, who can see the results of strategic open space protection, have made the campaign for publicly financed land preservation an easy sell in New Jersey.

Starting to Grow Smart in Somerset County

Located halfway between New York City and Philadelphia, the Fortune 500 mecca of Somerset County abuts Morris County's southern border, loosely aligned along the axis formed by the Watchung Mountains and the Raritan River. The home of major multinational firms like Upjohn, Johnson & Johnson, and AT&T, Somerset County consistently ranks among the wealthiest counties in the country, with a large component of white-collar affluence among its roughly 275,000 residents, found in places like Bernardsville, Gladstone, and Basking Ridge, and a fair measure of working-class boroughs such as Bound Brook, North Plainfield, Manville, Raritan, Somerville, and South Bound Brook. The county's wealth is largely a function of its growth, which, at 18 percent a year, is the state's fastest rate.

Less than two decades ago a semirural area, major roadways like Interstates 78 and 287 and U.S. Highways 1 and 202 have, according to a Somerset County official Denise Coyle, "changed the county," creating a landscape of "uncoordinated, unplanned development."[16] Malls, corporate headquarters, and congested roadways have become common in the county, extinguishing open space and farmland at a rapid clip. Since 1980, roughly half the county's farmland has been developed. Meanwhile, older downtowns in the county's six working-class boroughs are struggling, as economic flight has left behind six federal Superfund sites and close to three hundred other confirmed brownfields. Many are calling the situation dire. "There is a crisis. The crisis is overdevelopment. We're the fastest-growing everything. . . . This is the first time in 69 years that I'm considering moving out of the county because of my concerns with the quality of life," laments Ray Bateman, president of a local consulting firm.[17]

Comprising twenty-one municipalities, each with home-rule authority to make decisions and policies concerning development without regard to their potential negative effects on neighboring towns, Somerset has had to reckon with the problem of each municipality's fending for itself, saying yes to development because of the economic benefits but in the process undermining the area's overall environmental health and quality of life. Such development, which commonly occurs along the towns' borders, where the spillover effects of traffic, air pollution, and lost woodlands and open space on neighboring jurisdictions are most severe, has often led to confrontation between adjacent Somerset communities.

Municipal sovereignty is both a county and statewide planning problem. To address it, New Jersey became the first state in the country to adopt a sophisticated, bottom-up approach balancing the need to respect the home rule authority of municipalities with the necessity of coherent, consistent regional planning goals. In 1992, when the State Planning Commission and the Office of State Planning unveiled the State Development and Redevelopment Plan, the state's governing planning document, they devised a consensus-building process called cross-acceptance as a way of generating feedback on and support for the plan, thereby establishing planning policies from the ground up rather than imposing them from on high. Cross-acceptance involves a lengthy, multiyear process of discussion, revision, and compromise at every level of government and with numerous organizations and associations across the state. Each of the state's 21 counties and 567 municipalities must compare the plan with their own master plans and zoning laws and reconcile the differences.

The plan puts each municipality in one or more of five categories that determine the degree of development the state will support. Cities and older suburbs comprise the first two categories, defined by population density, development levels, infrastructure, and natural resources, with progressively less-developed areas comprising the other three categories.[18] Although the state does not have the power to enforce the plan, it can use its control of funds and its regulatory authority as leverage to encourage compliance. The plan manages predicted population and employment growth to the year 2010 by promoting "communities of place" and discouraging suburban sprawl. It calls for concentrating development in urban and nearby suburban areas and in "nodes" along major highways, where economies in infrastructure and public services can be achieved. The plan encourages clustering new development, using the state's rail system to revitalize downtown areas, and constructing mixed-use development of housing, stores, and offices accessible by a variety of travel modes—pedestrian, bicycle, train, and bus—rather than just the car.

With both the plan and the cross-acceptance process as a backdrop, Somerset County business and community leaders convened the first Somerset County Economic Summit in April 1997 to address the social, economic, and environmental problems brought about by unplanned growth and to establish a county-wide smart growth strategy. The summit participants articulated the challenge: "Somerset County has what most people want—a vigorous economy, a healthy environment, and vibrant communities. Goals generally pursued independent of the others. Somerset County also has traffic jams, suburban sprawl and vacant downtown stores—symbols of the uncoordinated and fragmented approach to development that threatens our economy and erodes our quality of life."[19]

With close to forty participants from area businesses, municipalities, county government, and civic and economic development organizations, the summit participants drafted a vision statement:

> The Somerset County we want to build is a place that our citizens and world class businesses choose to call home because of our success in preserving and enhancing our unique quality of life. Our engaged citizens have become national leaders in building collaborative partnerships—public and private, local and regional—to achieve this quality of life. We are a community that is dedicated to continued and sustainable economic vitality. We are dedicated to preserving our cultural heritage, embracing our growing diversity, protecting our natural beauty and open spaces, building vibrant town centers, and investing wisely in quality education, a highly skilled workforce, lifelong learning, affordable housing, and healthy community, and transportation.[20]

To achieve this purpose, the summit produced an action plan focused on preserving open space, revitalizing downtowns, building consensus and collaboration among municipalities, and retaining businesses. Specifically, the summit established five separate strategies, with short- and long-term goals, and assignments: (1) creating zoning and planning incentives to coordinate local government efforts better to achieve quality, smart development; (2) enhancing intergovernmental coordination in regulation to promote efficiency and quality development; (3) prioritizing county-wide funding to increase open space, revitalize town centers and neighborhoods, and create a regional center; (4) protecting the county's economic base through business retention, expansion, and targeted attraction; and (5) changing the mind-set and consciousness of county residents to create unity of purpose and a vision of a sustainable future. The summit also constituted the Stewardship Group to reach out to citizens and community organizations throughout the county to involve them in goal setting and refining the vision statement, and to enroll them as stakeholders in the implementation of the five smart growth strategies.

Less than a year later, in February 1998, under the leadership of freeholder Denise Coyle and several other elected officials and civic and business leaders, the second Economic Development Summit was held to evaluate progress on the county's smart growth strategies and to provide further input on implementation. Sixteen of the county's twenty-one mayors attended the second summit, as well as residents, elected officials, and representatives from nonprofit organizations, businesses, and educational institutions. Progress to that point included the establishment of a $2.4 million economic development revitalization fund to improve economic vitality in the county's six boroughs (including the redevelopment of an eleven-acre brownfield site in South Bound Brook, representing over 2 percent of the borough's land mass, that has been abandoned for over twenty years), passage of a three-cent county tax for open space acquisition, and the creation of a new, nonprofit smart growth organization, the Somerset Coalition for Smart Growth (SCSG), to help the county implement its smart growth strategies.[21]

The SCSG is the first organization of its kind anywhere, designed to work within the goals established by the State Development and Redevelopment Plan and provide technical assistance and advocacy services on behalf of Somerset municipalities and businesses on important smart growth techniques like brownfields redevelopment and downtown revitalization. As well, SCSG will build bridges between key county academic and civic organizations, such as the Chamber of Commerce, Raritan Valley Community College, and the Somerset Alliance for the Future, to promote collaboration and resource sharing.[22]

The second summit also produced a promotional smart growth video, *Somerset County: At a Turning Point,* as a tool to educate county residents about the problem of overdevelopment and build awareness of potential solutions. According to Denise Coyle, "Everyone knows there's problem, but not the scale or what can be done about it."[23] The video is intended to inspire action and imagination on the part of area residents, and to enforce the notion that at bottom the solution to the county's environmental and economic problems rests with each individual citizen, working together toward a common vision.

Still in its early phase, Somerset County's smart growth strategy has begun to gain momentum. But as Coyle points out, a lot of outreach and awareness building remain to be done. The core group of summit participants has succeeded in jump-starting the county's efforts, but it will take more than a handful of civic-minded leaders to put a stop to the development juggernaut and regain control of the county's destiny. The whole county, she enjoins, must rally to the cause.

Building a Blueprint for a Sustainable New Jersey

Barbara Lawrence, executive director of the nonprofit organization New Jersey Future, based in the capital city of Trenton, exudes enthusiasm and hope about the state's prospects for a healthy, livable future—what she calls a "sustainable state." "It's not too late for New Jersey. There are still plenty of choices to be made and the opportunity to make our hometowns more livable, and our economy and environment healthier."[24] Lawrence, who helped found Morris 2000 in 1984, has devoted most of her professional life to pursuing a strong economy, healthy environment, and just society for current and future generations, which happens to be New Jersey Future's mission. For Lawrence, a sustainable future means "aligning our activities so that we, and generations that follow, are able to maintain a high quality of life. To be sustainable, New Jersey must be able to view the critical trends that are shaping the future so that we can act on them with foresight and conscience."[25]

Leveraging the success and commitment of the dozens of New Jersey towns and counties engaged in open space protection and smart growth initiatives, and responding to the environmental mandate embodied in the 1992 state plan, Lawrence and her New Jersey Future colleagues started the Sustainable State Project (SSP) in 1995 to dispel the notion that economic expansion must come at the expense of environmental and social health. In developing the SSP, New Jersey Future looked to models of green planning from Europe and similar benchmarks from American cities and towns like Portland, Oregon, and realized that economic, environmental, and social goals must be set in tandem. Accordingly, the SSP's mission is to provide a forum for government officials, businesses, environmentalists, and civic leaders to work toward consensus on long-term, measurable quality-of-life goals and to create changes in New Jerseyans' behavior that help achieve those goals.

Since May 1995, New Jersey Future has sponsored an annual series of Leadership Conferences on indicators for sustainable development. With civic, business, and government leaders on hand, the conferences have attempted to set public goals for sustainability and devise measures for tracking progress toward those goals. Using the analogy of complex ecosystems maintained through intricate interrelationships that depend on well-developed feedback mechanisms, conference participants have worked to develop ways to generate information that would provide an accurate picture of the relative health and interconnectedness of the state's economic, environmental, and social systems.[26] Specifically, conference participants have focused on goals (broad statements that show what direction to take), indicators (measures of economic, social and environmental conditions that allow a comparison between where people want to be and where they actually are), and benchmarks (specific dates and quantitative measures, based on indicators, that are adopted through public process so as to inform participants about progress toward our goals). As an example, "fishable, swimmable rivers" is a goal; actual measurements of dissolved oxygen in the rivers are indicators; and a level of 4.0 parts per million of dissolved oxygen in 50 percent of New Jersey's rivers by the year 2010 is a benchmark.

The key to establishing good, reliable goals, indicators, and benchmarks, according to Lawrence, is a public process that engages a wide range of civic and advocacy organizations as well as the public at large. "Though they may seem technical—and must indeed be backed by sound, though sometimes inscrutable, science—goals, indicators and benchmarks," Lawrence stresses, "are based on value assumptions; citizens must be fully involved in deciding those values. . . . Once they are provided with adequate information, citizens—individually and through their civic associations—can see beyond government institutions toward new relationships, and conceive new pathways to their future well-being."[27]

Another critical function of indicators of progress is that they enable citizens to be more knowledgeable about the collective consequences of their individual actions. As Lawrence explains, "This knowledge is essential to voluntary change, and to effective leadership. The engagement of individual citizens is necessary, first, in deciding on responsible practices, and then in carrying them out. Reductions in air, water, and land pollution and in the consumption of resources must begin at the source, very often with household practices; lawn care, waste disposal, automobile travel."[28]

The environmental indicators New Jersey Future has helped to develop, spelled out in the 1999 Sustainable State Report, include the following:

1. **Efficient Transportation and Land Use.** The goal is to create a choice of efficient, convenient, safe and affordable transportation and land use options that provide access to jobs, shopping, recreational centers, schools, airports and rail centers. Indicators include:
   A. Backlog of infrastructure repairs: This measure shows how well the state is doing at maintaining existing transportation infrastructure, which is essential for a strong economy to provide convenient access to many necessities of daily life.
   B. Vehicle miles traveled and public transit ridership: This is a measure of auto-dependency and how many people use New Jersey's mass transit system, which is more efficient in terms of land consumption and less polluting than automobile travel.
   C. Transportation choice: This measure reports on the transportation choices available for the 10 largest office developments and fastest growing residential areas.
   D. Pedestrian and auto-related deaths: This is an indicator of how well communities are planned and designed to provide for safe pedestrian access, of how much and how safely New Jerseyans drive, and of suburban sprawl.
2. **Natural and Ecological Integrity.** The goal is to preserve and restore New Jersey's ecosystems and the full complement of wildlife and fish species that share the state with humans. Indicators include:
   A. Freshwater wetlands lost per year: Wetlands are essential for filtering pollutants from drinking water, preventing floods, and as habitat for native wildlife.
   B. Nesting waterbird populations: Every year great numbers of waterbirds gather near the New Jersey coast to breed. Habitat loss affects these numbers.
   C. River health: The presence of dissolved oxygen in streams is an important indicator of the overall health of river ecosystems.
   D. Marine water quality and shellfish: This is measure of the percent of shellfisheries not closed due to pollution.
3. **Protected Natural Resources.** The goal is to maintain New Jersey's natural resource base. Indicators include:

A. Energy consumption per capita: This measure shows how much energy an average resident uses in one year. Energy is perhaps the most limited resource, and the state's economy and quality of life depend on it.
B. Acres of farmland: Like energy, arable land for food production is essential to the state's economy and quality of life. This indicator tells how much land remains for agriculture and other food production enterprises.
C. Beach and bay closings: This is a measure of the number of times a beach or bay had to be closed due to unsafe environmental conditions.
D. Preserved and developed land: This tells how many acres of land have been preserved or developed.

4. **Minimal Pollution and Waste.** The goal is to minimize the generation and accumulation of pollution and waste, maximize the use of efficient, clean and sustainable energy sources, and to increase consumer choices of ecologically-friendly products and services. Indicators include:
   A. Drinking water quality: This measure reports how well water systems are providing clean drinking water.
   B. Greenhouse gas releases: Fossil fuel use, especially from automobiles, is the major source of greenhouse emissions in New Jersey and must therefore be reduced.
   C. Total solid waste production per capita: This is a measure of how much garbage each New Jerseyan produces in a year. Garbage disposal, by landfills or incineration, is a major source of air, water and soil pollution.
   D. Air Pollution: The presence of ozone, air borne particles, and carbon monoxide are important indicators of health and air quality, as reflected in the annual number of unhealthful air days.

Other nonenvironmental goals, each with its own set of indicators, include "Strong Community, Culture, and Recreation," "Economic Vitality," "Quality Education," "Equity," "Good Government," "Decent Housing," and "Healthy People."

These goals and indicators have become the new yardstick with which New Jerseyans measure their progress, replacing traditional, inadequate measures like the gross state product, the state analogue to the national gross domestic product, which do not account for social values such as human health, environmental quality, social justice, and other elements of genuine sustainability.

In developing its Goals and Indicators for a Sustainable State, New Jersey Future identified five key challenges in achieving sustainability, along with potential solutions:

1. *Accounting for long-term and large-scale impacts and costs in present time.* Individuals, businesses, and governments have an incentive to shift costs away to another place and time, which economists call "externalizing" costs. These costs do not disappear, but instead show up as environmental harms, social inequities, and the like. In New Jersey, home-rule authority and over-reliance on property taxes force municipalities to compete for development to balance budgets, while externalizing costs like affordable housing or spillover traffic in adjoining towns. Regional planning and coordinating institutions can help meet this challenge.
2. *Facilitating more widespread understanding of complex issues.* Many ordinary citizens and public officials often lack the specialized or technical information that is essential to informed participation and good public policy decisionmaking. Further, information that enters public debate is often colored by special interest groups with a particular stake in the outcome. New ways must be created to frame complex questions for effective, democratic debate.
3. *Resolving the conflict between interrelated public goals and fragmented governmental responsibilities.* Governmental decisionmaking is fragmented, both horizontally among agencies of the same sovereign, and vertically between federal, state, and local governments. For example, despite the overlapping responsibilities of health, environmental, transportation, and land use agencies, there has never been an effective mechanism for linking them. Agencies are thus free to work at cross-purposes, with different values, and against the long-term interests of society. Broad-based, goals-oriented processes are necessary to help public agencies and civic organizations integrate and amend their missions.
4. *Bridging the gap between public plans and government operations.* Planning without teeth can lead to unsatisfactory results. For example, the State Planning Commission lacks the statutory authority to enforce the State Plan, thus leaving open the question of whether the plan will be implemented effectively. Planning must be enforceable.
5. *Meeting the challenge of meaningful civic engagement.* While mechanisms like the state's cross-acceptance process represent an important step toward ensuring civic engagement in land use decisionmaking, it is only a beginning. Citizens must be involved in every social policy decision, especially those who traditionally have not had a pubic voice, with reliable information at their disposal.

These challenges are the operational hurdles that must be surmounted to implement the SSP fully. As Barbara Lawrence states, "If New Jersey's public and private leaders address these institutional challenges with determination—through civic action, agency reform, executive order, or legislation—sustainable development will become a reality in New Jersey, and we could enter the twenty-first century with a new level of confidence."[29]

In 1997, New Jersey Future persuaded Governor Christine Todd Whitman to adopt the SSP as the cornerstone of her administration's sustainable development strategy. With great fanfare, the governor announced the SSP in the inaugural address of her second term in January 1998:

When I speak about strong cities and bike paths and open space, I'm talking about a sustainable society—a society in which we protect the resources we have today so they are there for us tomorrow. We all have a stake in this, and we need to know where we stand.

So, for the first time, the State of New Jersey will establish ways to measure our quality of life and report on our progress. We'll let you know about things that matter to you: the quality of our water, the status of our cities, the traffic on our streets, the health of our children. And we'll tell you just how we're doing in saving open space and containing sprawl—good, bad, or ugly.

Together, we can create a New Jersey we will be proud to pass on to our children and grandchildren. A New Jersey in which all our communities prosper, in which fertile farms, sparkling waters, and breathtaking mountain views remain lasting treasures.[30]

Underlying the rhetoric befitting an inaugural address is real substance, Lawrence says. The Whitman administration, she claims, has fully bought into the SSP, becoming a driving force. "The support of the Governor has been essential to the success of the [Sustainable State] Project to date," she emphasizes.

Yet the bulk of the work remains with New Jersey Future and the citizenry of New Jersey. This is, after all, not a top-down effort but a historic civic-led experiment in statewide environmental and social action. "As we enter the new millennium, New Jersey is at a remarkable historical crossroads," Lawrence describes:

We are in the midst of the best economic period in recent memory, yet we as a society face many critical questions about the future. Not only questions about our immediate future here in New Jersey, but questions about our long-term collective futures as residents of planet earth. Through the Sustainable State Project, we will attempt to define what is important so that we may look ahead knowing some of the fundamental conditions that will shape our collective future. If we can ask the right questions, and work to answer them, we can look to the twenty-first century with renewed confidence.[31]

In addition to developing and institutionalizing long-term, strategic goals and benchmarks and creating an ongoing public process with leadership from the governor, the SSP is developing special projects to build new alliances among government, businesses, and environmentalists in the state. The Green and Gold Task Force, for instance, brings together a diverse group of business, conservation, civic, and local government representatives to assist the New Jersey Department of Environmental Protection in deciding how best to meet its obligations under the state plan.

Ultimately the SSP is an attempt to build a new system of governance based on comprehensive, systems-oriented planning that will guide public policy and private actions in accordance with the SSP's goals and benchmarks. Planning is thus the process for identifying goals and the steps to achieve them, and around which there is public consensus. Robert Currie, former director of strategic planning and management at the U.S. EPA and a Leadership Conference participant, explains that planning and plan implementation move beyond a reactive, regulatory model of governance. Traditional environmental policymaking, for instance, is based on the "you break it, then you fix it" approach, which, he argues, is like "rowing a boat, looking backward all the time—it is not clear where you are going. We are reacting to problems; we are not planning. We are not managing our future."[32]

Further, a planning-based approach to social and environmental policy necessarily transforms regulation from a top-down, command-and-control approach to one that emphasizes a systems approach as well as leadership and responsibility on the part of individuals, communities, and businesses. In environmental terms this means encouraging these stakeholders to assume greater responsibility in meeting environmental goals and to look at the entire system of production and consumption, leading to a pollution-prevention strategy focusing on ecosystems as a whole rather than specific impacts.

Implicit in this model is the notion that civic engagement—the participation and responsibility of individuals, communities, and firms—is essential to achieving environmental results. Recognizing that individual and community actions have a significant impact on environmental quality, including vehicle miles traveled, household consumption, and lawn care, for example, these stakeholders, and not just government and businesses, must be brought to the table. Vernice Miller, another conference participant and leading environmental justice activist, states, "Communities have to be involved in environmental decisionmaking and policymaking; there has to be real public participation, not nominal participation, where communities are part of the goal setting and the agenda setting." Such participation, she added, must be "culturally reflective of who you are as a state."[33]

Moreover, Miller urges, goal setting and benchmarking must be performed with racial and economic equality in mind; otherwise the work of the SSP will be fatally undermined. She warns that "every community has to be viable and every community has to be environmentally-sound in order for the state to move forward to achieve its long-term objectives. Once we begin to grapple . . . with making those communities whole, the state is going to be able to move forward. We can't if some of us are inherently living in unequal situations."[34]

The SSP also reflects a regional approach to the state's environmental, economic, and social problems exemplified by the companion problems of sprawl and urban disinvestment. The "race-to-the-bottom," parochial behavior often encouraged by home rule authority and the absence of comprehensive regional planning undermines smart growth and sustainability goals, which are by their very nature regional in scope. Regionalism knits the concerns of urban, suburban, and rural constituencies around the common goal of healthy, livable communities and establishes governance mechanisms designed to address regional issues like open space preservation, traffic congestion and mass transit, urban revitalization, and racial segregation. Leading regionalist advocates like Minnesota congressman Myron Orfield have highlighted the causal relation between sprawl and fragmented governance. Orfield coined the term *metropolitics* to describe the new forms of governance and policy necessary to combat sprawl and urban disinvestment.[35] Regionalists point to success stories in Portland, Oregon, and Toronto, Canada, where regional governance models have prevented zero-sum competition for development among local jurisdictions, coordinated zoning and transit initiatives to match regional needs, and streamlined infrastructure improvements to avoid redundant capital investments by municipalities. The SSP attempts to replicate these successes in regional governance in New Jersey.

Under the SSP model, therefore, environmental policy, and all social policy, must be integrated within a planning framework, informed by the interconnections that exist among natural and human-made systems, reflective of the regional scale of environmental, social, and economic issues, and committed to social and environmental justice. Regulation, under this policy approach, is merely a technique—a means by which plans are implemented.

With four years of work and the governor's commitment, New Jersey Future has rolled out the Goals and Indicators for a Sustainable State and begun the process of engaging citizens, civic associations, municipalities, and businesses from across the state in the effort. Complementing the open space and smart growth initiatives that have blossomed in dozens of New Jersey counties and towns in the 1990s, the SSP has begun to synthesize these and related efforts into a statewide movement for sustainability and is on its way to changing the consciousness and behavior of New Jerseyans.

Civic Environmentalism and New Jersey's Sustainability Efforts

That New Jersey provides such compelling case studies in civic environmental action should not be surprising. After all, necessity is the mother of invention. The pervasive and persistent environmental degradation that has brought the state so much notoriety has also motivated countless citizens, civic organizations, businesses, and governments across the state to engage in innovative experiments in social action. The response to the state's physical decline has served as the occasion to examine New Jersey's overall quality of life and to redefine the terms by which the state measures its progress.

From the municipal level up to the governor's office, a diverse group of stakeholders has undertaken innovative planning measures designed to mesh environmental, social, and economic goals under the headings of smart growth and sustainability. Part and parcel of these planning-based strategies is citizen involvement and education, as evidenced by efforts such as Randolph Township's Open Space Committee, Somerset County's smart growth video, and New Jersey Future's Leadership Conference series.

Consistently, leaders of the New Jersey initiatives have expressed the importance of outreach and education to the success of their efforts, given the causal connection between individual and private actions and many of the state's most pressing environmental harms.

Further, the smart growth and sustainability projects include a commitment to promoting social and environmental justice goals like affordable housing for lower-income residents and brownfields cleanup and redevelopment in hard-hit urban neighborhoods. In the light of their comprehensive planning approach and commitment to sustainability in its fullest sense, the New Jersey efforts do not make distinctions among environmental, social, and economic goals. Instead, they are viewed as inextricably linked.

The SSP's focus on pollution-prevention goals and redefinition of economic growth to incorporate environmental and social costs is an example of industrial ecology principles in action. On the ecoindustrial model, economic growth does not come at the expense of environmental health or overall quality of life; rather it protects and preserves natural capital for the benefit of current and future generations. Ecosystems provide the living analogy with which the SSP has been designed, and ecosystem health, including the viability of human communities, is the ultimate yardstick of sustainability. Whether at the organizational level or statewide, industrial ecology is at the heart of the SSP's mission.

Finally, as articulated in the State Development and Redevelopment Plan, New Jerseyans seek to create and restore "communities of place." Place has been under assault in New Jersey because of overdevelopment, as highways, subdivisions, and shopping malls have robbed entire communities of much of their natural beauty. The writer James Kunstler calls these places "the geography of nowhere."[36] Recognizing this psychic and physical loss, citizens from the state's cities, suburbs, and rural townships have banded together in local, county, and statewide efforts to redeem the Garden State and bring back the environmental quality that name implies.

The road ahead is long, and paved with good intentions that will be realized only if the leadership and energy of citizens in places like Morris and Somerset counties and in organizations like New Jersey Future trickle down throughout the citizenry. The new model of governance based on comprehensive regional planning and civic participation, embodied in the SSP's Goals and Indicators for a Sustainable State, depends on an active and diverse constituency committed to the common good in its most concrete terms. This means that New Jerseyans must generate the civic will that can thwart shortsighted decision-making based on parochial motives and private gain and institute long-term social policy choices grounded in a concern for the overall social, economic, and environmental health of New Jersey's communities.

In addition, this new governance structure must integrate and coordinate the planning of many different jurisdictions, from the municipal and county levels to the state, regional, and national levels. Systems-oriented problems demand systems-oriented solutions, including the cooperation of governments so as to ensure they are not working at cross-purposes and instead are able to design policy solutions that match the scope of the problem. Governmental cooperation also helps protect against unwanted or unintended consequences by providing open channels of communication to identify and respond to problems early and by creating accountability among municipalities and counties that might otherwise try to externalize social and environmental costs.

Further, to the extent that smart growth strategies, especially open space preservation and brownfields redevelopment, succeed in channeling development activities to urban or already built-up areas, New Jersey governments must adequately invest in infrastructure improvements and housing to accommodate this new growth. Taking a systems approach to planning means anticipating the redistributive effects of managing growth and dedicating resources accordingly. Municipal, county, state, and federal funds must be made available for the kinds of capital improvements necessary to sustain targeted growth.

Likewise, New Jersey Future, government agencies, and others spearheading New Jersey's smart growth and sustainability efforts must aggressively reach out to urban constituencies, especially in hard-hit inner-city neighborhoods in places like Camden, Trenton, and Newark, to ensure their participation. Much of the success of any smart growth strategy hinges on the capacity of urban areas to be involved in and take advantage of environmental and economic development decision making. Urban constituencies must be able to speak for themselves and control their own destiny, in cooperation with their suburban and rural counterparts. This is what smart growth and sustainability are all about.

It must also be understood that simply purchasing open space with taxpayer dollars is the easy way out. It is much more difficult to undertake comprehensive planning that addresses both the built environment and undeveloped land. To the extent that states and municipalities buy up islands of land while leaving unaddressed the more complicated problem of how to build and develop in an ecologically sound manner, they will have achieved only a small success. Dollars for land acquisition are one thing; comprehensive, long-term planning for sustainability is another.

Further, open space acquisition itself is not necessarily an environmental boon. The very term *open space* can be interpreted as describing an empty place, a simple quantum of land devoid of quality habitat or natural features. Open space can mean almost anything: a prison yard, a golf course, a landfill, or an abandoned parking lot. Accordingly, data regarding the amount of open space protected in any particular community or jurisdiction can be misleading. How much open space in New Jersey actually satisfies a more rigorous standard for environmental quality?

Forty-five percent of the total land area of natural ecosystems in the United States has been converted into cropland, pasture, plantations, or human settlements in the nearly four centuries since the first European settlers came to these shores.[37] That is a lot of open space created as a direct result of ecosystem destruction.

In addition, open space acquisition does not necessarily address the critical problem of endangered ecosystems, especially plant life. According to the first comprehensive assessment of plant endangerment in the United States conducted by the World Conservation Union, one out of every three plant species is under threat of extinction. With 4,669 of its plant species threatened by agriculture, logging, development, and invasion by foreign plants, the United States ranks first, by far, among all nations in total number of plants at risk. In all, 29 percent of the United States' 16,108 plant species face extinction.[38] More fundamental to human life and the overall functioning of nature than mammals and birds, plants support most of the rest of living organisms by converting sunlight into food. They provide the raw material for a multitude of medicines and agricultural products, and constitute the bulk of the physical structure of the landscape.

Open space acquisition, although an important environmental protection technique, by no means guarantees the preservation of healthy ecosystems. Perhaps a better, more ambitious goal than open space protection is the preservation and restoration of plant and wildlife habitat: the places that sustain a diversity of life forms. In every state, environmentalists and citizens alike should aim to protect and restore areas that function as genuine habitat for wildlife, plants, and other essential ingredients of ecosystems, and not merely empty stretches of land. The quality of the open space preserved should be just as important as the quantity. Open space indicators should differentiate between the amount of habitat versus the amount of space acquired.

Finally, there is the controversial question of growth, smart or otherwise. Many environmentalists such as Donella Meadows and Bill McKibben have long argued that Americans reached the ecological limits of growth some time ago and continue to push the bounds of natural carrying capacity well beyond a safe threshold.[39] Meadows explains that smart growth "reinforces the view that growth will occur anyway, that we can't stop it, and that it's good. Calling it 'smart growth' misses the point—we're already too big. We must un-grow. The single most important thing we who care about the environment can do is throw ourselves at our deeply held, cultural assumption of growth, and stop paying lip service to what was a bad idea in the first place."[40] McKibben echoes this sentiment, claiming that human society "may already be larger than what the planet can sustain in the long run. We should be asking ourselves whether we'd be better off with smaller populations and appetites."[41]

To many, New Jersey is proof of this very point. Population, development, and pollution have already gone too far. According to critics like Meadows and McKibben, "smart growth" is nothing more than a contradiction in terms, a fancy policy slogan signifying nothing and misleading the public into believing that growth can be effectively managed.

"No-growth" advocates like Meadows and McKibben have an important message to send, especially to a culture at the height of its economic powers and consumed with the pursuit of higher stock prices, bigger cars, and ever larger homes. Yet considering the economic realities that most Americans face, especially those at the low end of the income scale struggling to find or, at best, keep a job, economic growth is a political and social necessity. Environmentalists and others cannot responsibly look into the eyes of un- or underemployed Americans and say, "Sorry, you can't have a job because the planet's too small." Economic and social justice dictate that environmentalists find a middle ground.

Thus, insofar as smart growth strategies direct economic growth to places such as inner-city neighborhoods where quality job opportunities are scarce, while prohibiting the same in areas where the need is less severe, growth can be seen as a positive force, assuming it does not come at the expense of local environmental quality. To be "smart," growth must make sense not only in terms of where it occurs but how it occurs, which is what sustainable development is all about. Sustainable development implies social equity and environmental quality, ensuring the development of healthy human communities in perpetuity. In this sense, what is wrong with the idea of smart growth is the idea of growth itself. It is a bald term, denoting quantity, not quality. Environmentalists should thus insist that we reinvent the idea of growth so as to ensure that environmental quality and social health are part of its meaning, which is what the SSP is attempting to do.

New Jerseyans are not alone in their attempt to reinvent their governance structure and the goals by which they measure their success. Towns, cities, and counties across the country are beginning to do the same. But in the light of the sheer magnitude of New Jersey's environmental problems, its success so far in developing smart growth strategies and reaching out to citizens from diverse communities, New Jersey is uniquely poised to set the standard for sustainability. If New Jerseyans can muster the civic will, they are bound to succeed.

8

Coming Full Circle: An Emerging Model of Environmentalism and Democracy

We need to embrace the full continuum of a natural landscape that is also cultural, in which the city, the suburb, the pastoral, and the wild each has its proper place, which we permit ourselves to celebrate without needlessly denigrating the others. . . . In particular, we need to discover a common middle ground in which all of these things, from the city to the wilderness, can somehow be encompassed in the word "home." Home, after all, is the place where finally we make our living. It is the place for which we take responsibility, the place we try to sustain so we can pass on what is best in it (and in ourselves) to our children.

—William Cronon, *Uncommon Ground*

The final test of an economic system is not the tons of iron, the tanks of oil, or the miles of textiles it produces: The final test lies in its ultimate products—the sort of men and women it nurtures and the order and beauty and sanity of their communities.

—Lewis Mumford, *Faith for Living*

Pursuing Environmentalism and Democracy

From Massachusetts to California, Colorado to New Jersey, a diverse set of communities are pursuing environmentalism and democracy. Confronted with a challenging mix of social, economic, and environmental problems that undermine their quality of life, sense of place and community, and physical well-being, these and many other communities are responding in creative, even visionary ways, working against great odds to take charge of their destinies.

Changes in the American landscape, in its physical and spatial order, reflect discrete social and economic changes, the result of the fundamental dialectic between environment and culture. Environmental harms cut across geographic boundaries and are both a cause and an effect of adverse societal conditions like declining social capital, political disaffection, rising economic inequality, racial segregation, and excessive privatization.

Increasingly communities are transforming their environmental problems into an opportunity to address underlying social and economic problems, and they are using the lever of environmental protection strategies to push for systemic changes in the way people relate to each other and to their environment. As the four case studies demonstrate, the immediacy, persistence, and pervasiveness of environmental degradation spur social action and lend themselves to systems-oriented solutions. This is the great lesson of these case studies, and of environmentalism itself.

Civic environmentalism is the emerging model of social and environmental activism. It is a dynamic and transformative enterprise that moves beyond top-down, centralized law and regulation to planning and implementation at the community and regional levels. It embraces an ecosystem approach to social problem solving, with the environment as both a prime subject and a principal metaphor of civic action. Civic environmentalism does not just focus on specific media or pollutants, as traditional environmental regulation does. Rather, it focuses on the overall health and quality of life of communities—social, economic, and environmental—and the sustainability of that health and quality of life over time. Personal relationships and networks, neighborhoods and livelihoods, across geographic, cultural, and political borders, are the fuel that powers civic environmental projects and give rise to strategies backed by social investment, such as open space taxes, brownfield cleanups, conservation easements, and mass transit. Civic environmentalism links urban, suburban, and rural constituencies in the pursuit of shared goals and visions, and enforces the notion that our fates are bound together by place and time.

Civic environmental strategies take different forms: urban agriculture and brownfields redevelopment; transit-based development and waterfront restoration; conservation-based development and rural land preservation; suburban open space protection, smart growth planning, and indicators of sustainability. These strategies suggest a realignment of environmental protection strategies away from a purely law-based, law-driven, and professional model to one in which a diverse group of citizens, environmentalists, government officials, and businesspeople collectively become the experts and implementers. Civic environmentalism acknowledges that just as the bulk of our pollution problems stem from sources other than smokestacks and outflow pipes and that conventional pollution control strategies are often a zero-sum game, so too must responsibility for dealing with environmental degradation be a collaborative enterprise involving all the stakeholders who inhabit a shared space, be it a blighted inner-city neighborhood, a neglected urban waterfront, a rural county under siege from development, or a congested, sprawling suburb. Without doing away with the important and powerful stick of environmental enforcement and public interest litigation, civic environmental strategies help close the loop by allowing environmentalists and other stakeholders to move beyond an adversarial approach to one in which economic development and environmental protection are part of the same goal.

Civic environmentalism is becoming the trend among mainstream environmental organizations. For example, the Sierra Club and National Audubon Society are retooling themselves, shedding their centralized structure and returning to a chapter- and locally based organizational model while focusing on land use and development issues like sprawl and ecosystem protection. Meanwhile, groups like the Nature Conservancy and Trust for Public Land are expanding their toolbox of land conservation strategies, emphasizing approaches that increase civic involvement and protect working landscapes, whether in inner-city neighborhoods or distant rural communities. New England's Conservation Law Foundation (CLF) has recently formed local advisory committees for each of its four regional offices to engage citizens more directly in decision making and programs. At the same time, CLF is augmenting its advocacy and litigation programs with a new environmental ventures agenda, helping to launch environmentally responsible development projects and products.

Ironically, development, at the root of so much environmental damage, has become central to many civic environmental solutions in the form of brownfields redevelopment, transit- and conservation-based development, and ecoindustrial parks. Such green development strategies, and the concept of industrial ecology in general, have provided environmentalists with powerful tools to influence development decision making and redefine what economic growth and progress mean. Meanwhile, just as conservation groups are transforming themselves into enthusiastic promoters of working landscapes, agricultural enterprises, and other conservation-based development projects, community organizations, including community development corporations that traditionally only dabbled in environmental projects, are becoming environmental leaders.

At the heart of civic environmentalism are six core concepts that provide a solid theoretical framework:

1. *Democratic process*—engaging a critical mass of stakeholders united by place and time in the project of rebuilding and restoring communities' social, economic, and physical fabric
2. *Community and regional planning*—integrating law, science, fiscal policy, governance mechanisms, and civic will in systems-oriented environmental strategies
3. *Education*—building greater awareness among citizens of the environmental and social costs of their actions ("environmental literacy") and enlisting them in the process of identifying and remedying environmental harms
4. *Environmental justice*—embedding in all environmental strategies a concern for lower-income people and communities of color, who have been disproportionately burdened by environmental harms in the past and must be directly engaged in environmental decision making
5. *Industrial ecology*—the principles of pollution prevention and sustainability as applied to the design and production methods of firms and to economic development generally and the theoretical framework for what eco-architect Bill McDonough and others are calling the next industrial revolution
6. *Place*—the physical and emotional space we inhabit that inspires civic action

These core concepts explain the thrust and direction of civic environmentalism, bridging the gap between environmental protection, economic development, and community building. They comprise the cornerstone of a new kind of environmental and social action, aimed at achieving livable, sustainable communities for all.

The Challenges to Civic Environmentalism

Despite the promise of civic environmentalism, several challenges stand in the way of large-scale social and environmental progress. The forces of global capitalism—the people and markets that control the flow of capital across place and time—continue to externalize social and environmental costs, both domestically and around the globe, in the pursuit of short-term profits. These same forces are undermining people's commitments to civic life and place by physically displacing and distracting them from social problem solving through sanitized, corporatized media and entertainment.

The emergence of the Internet poses further potential problems for civic environmentalism. Pervading every aspect of private and public life, the Internet and electronic technologies in general have begun to transform the way we conceive ourselves as communities and citizens. As much as the Internet has facilitated meaningful communication and relationships among users and is an important tool for gathering and disseminating information, its primary power has come from its commercial use, which one writer has perversely called a "new commercial ecosystem." Further playing on this environmental metaphor, a leading Internet investment banker has even gone so far as to describe the frenzy to invest in Internet companies as a "land-grab."

Like other "land grabs" in American history, this one threatens to undermine the development of a durable civic discourse and consequently civic environmental strategies. By consuming citizens' time and energy with what amounts to nothing more than an around-the-clock, placeless cybershopping mall, dominated by supercompanies with flashy web sites and hip advertising campaigns, the Internet is widening the already capacious distance between citizens and the real, tangible places they inhabit yet too easily ignore. It is further undermining the experience of place with the experience of placelessness and anomie, and replacing meaningful association between people and their environment with nonstop electronic interactions between firms and consumers.

So much money and media attention are directed to the Internet, aggrandizing and even fetishizing it, that less commercially compelling, though far more important stories like the success of civic environmental projects risk being marginalized or entirely overlooked. If only a fraction of the creativity, time, and money that go into launching and operating Internet companies were devoted to constructing and restoring healthy communities, we all would be the better for it. Instead, as Aldo Leopold lamented over a half-century ago, we dedicate the vast majority of our energies to building a "better motor," or in this case, software or web site, yet throw dice when it comes to building the places where we actually live and work.

Another stumbling block is excessive privatization, reflected in the uncompromising defense of private property interests and the flight of the burgeoning professional class to gated enclaves. Privatization of this sort prohibits communities from realizing public goals. Open space protection requires a commitment to public funding and community planning as well as a willingness on the part of private property owners to sacrifice short-term gains in the interest of long-term, intergenerational community benefits. Similarly, significant reductions in air pollution levels demand public investment in mass transit, among other public-spirited measures. In one way or another, every environmental protection strategy requires a conception of the common good, of the public interest, stemming from our shared dependence on the air, water, and soil. When we lose that sense of a common destiny, we lose our ability to act on it.

Yet another challenge confronting civic environmentalism is the decline of public discourse. In an age when essentially private matters—whether the rise and fall of the Dow Jones Industrial Average, political sex scandals, or the salaries of celebrity athletes—seem to dominate the airwaves and newspaper headlines, it is difficult to generate a sustained, meaningful conversation about public issues like environmental quality, public education, health care, or the cost of living. Lacking such a discourse, civic will languishes or is extinguished altogether, making it impossible to plan, organize, and execute public strategies. In the robust discussion of public problems are the seeds of the solutions. Environmental protection, like all other public projects, is as much a matter of civic commitment as it is a matter of law and policy.

Yet law and policy matter a lot. Simply relying on the goodwill and voluntary commitments of citizens, businesses, and others is a risky strategy. If civic environmentalism is to prosper, it will require the establishment of laws and policies that enforce civic environmental goals, such as pollution prevention at the firm level, or regional mass transit systems, or restoration of degraded habitat, and ensure both citizen participation and accountability at all levels of business and government. Law and community, like the environment and community, constitute one another. If done right, they can benefit each other.

We do not necessarily need more laws; rather, we need new and better laws—laws that do not simply control pollution from individual sources or redistribute it, but that provide incentives and demand the elimination of pollution altogether. We need laws that require government agencies and businesses to consult with and listen to each other and community stakeholders when developing projects with regional environmental and social impacts such as highways, factories, office parks, or housing. And we need laws that make it easier for individuals and businesses to make environmentally responsible decisions, whether about where to live, what products to buy or sell, or what form of transportation to take. Law is the public expression of a community's values and goals. It should support them, not undermine them.

Repairing the Fabric, Redeeming the Land

Still in its early stages, civic environmentalism has a long way to go in achieving its ambitious goals. Nevertheless, civic environmental strategies are generating momentum and hope in communities across the country—in places like Roxbury, Massachusetts; Oakland, California; Clark, Colorado; and Randolph Township, New Jersey. The planning activities these communities have undertaken are major steps toward realizing sustainable communities. After all, good planning is the necessary first step in any civic environmental strategy; planning is action in the fullest sense.

Civic environmental projects are tapping into the public's pent-up demand for effective, hands-on community-building strategies. They are providing unique opportunities for people from different communities and walks of life to work together toward the goals that, at bottom, all people share: a healthy environment, quality jobs, a sense of place and community. But opportunity alone is not enough. The challenge ahead is to build on these successes and create a new standard of environmental protection, economic development, and community building, a new standard of sustainable development.

At the beginning of this book are excerpts from two very different writers, of different epochs and cultures, who capture the spirit of the new environmental age dawning at the beginning of the twenty-first century. They are the visionary nineteenth-century Vermont conservationist and writer George Perkins Marsh and the great folk poet of post–World War II Harlem, Langston Hughes. In his seminal work of 1864, *Man and Nature*,

Marsh wrote: "In reclaiming and reoccupying lands laid waste by human improvidence or malice, the task is to become a co-worker with nature in the reconstruction of the damaged fabric."[2]

Roughly seventy-five years later, Hughes penned these lines in the poem "Let America Be America Again":

O, let America be America again
The land that never has been yet
And yet must be—the land where every man is free
. . . We, the people, must redeem
The land, the mines, the plants, the rivers
The mountains and the endless plain
All, all the stretch of these great green states
And make America again![3]

Coming full circle, an emerging civic environmentalism marries these two American visions, creating a whole equal to the task of rebuilding America's communities from the ground up, of repairing the damaged social and environmental fabric that is American society at the beginning of the twenty-first century. It unites Marsh's vision with Hughes's in the project of becoming coworkers with nature, to make America again, fulfilling environmentalism's promise as democracy's boldest symbol and practice.

Notes

Preface

1. D. Hayden, *The Power of Place: Urban Landscapes as Public History* (Cambridge: MIT Press, 1995), 9.

2. S. Schama, *Landscape and Memory* (New York: Random House, 1995), 10.

3. Hayden, *Power of Place,* 19.

4. D. Kemmis, *Community and the Politics of Place* (Norman, OK: University of Oklahoma Press, 1990), 118.

5. B. Jordan, quoted on the WGBH television program "Say Brother."

6. G. Lakoff and M. Johnson, *Metaphors We Live By* (Chicago: University of Chicago Press, 1980), 5.

7. B. Barber, *Strong Democracy* (Berkeley: University of California Press, 1984).

Introduction

1. "Cultivators of the earth are the most valuable citizens," Thomas Jefferson proclaimed, "They are the most vigorous, the most independent, the most virtuous, and they are tied to their country and wedded to its liberty and interests, by the most lasting bonds." T. Jefferson to John Jay, Aug. 23, 1785, in J. P. Boyd, ed., *The Papers of Thomas Jefferson* (Princeton: Princeton University Press, 1952), 8:426.

Notwithstanding what Americans today now rightly understand to be the hypocrisy and inadequacy of Jefferson's social vision, shot through with racism, cultural imperialism, and an antiurban stance, Jefferson, methodically driven by his dream of a nation of virtuous yeoman farmers, shaped the trajectory of American history like no other statesman and bequeathed to the nation an ideological legacy that still colors our collective consciousness.

2. The cultural historian Jackson Lears has written recently about the curious rise of nostalgia in American historiography. While taking to task those historians who would blindly and unscrupulously embrace the past as pristine and virtuous, Lears cautions against dismissing the past as fatally flawed and devoid of value as a guide

to living in the present. While in no way a story of declension, Lears suggests that American history reveals the erosion of certain social and cultural values associated with community and respect. Nostalgia, he writes, can thus be an "energizing impulse, maybe even a form of knowledge. The effort to revalue what has been lost can motivate serious historical inquiry; it can also cast a powerful light on the present. Visions of the good society can come from recollections and reconstructions of the past, not only from fantasies of the future." "Looking Backward: In Defense of Nostalgia," *Lingua Franca* 7 (Dec.–Jan. 1998): 66.

3. See R. Putnam, "Bowling Alone: America's Declining Social Capital," *Journal of Democracy* 6 (Jan. 1995); C. Lasch, *The Revolt of the Elites and the Betrayal of Democracy* (New York: Norton, 1995); M. Sandel, *Democracy's Discontent: America in Search of a Public Philosophy* (Cambridge: Harvard University Press, 1996)

4. See Sandel, *Democracy's Discontent*; G. Wills, "Washington Is Not Where It's At," *New York Times Magazine*, Jan. 25, 1998.

5. C. Cobb, T. Halstead, and J. Rowe, "If the GDP Is Up, Why Is America Down?" *Atlantic Monthly* (Oct. 1995).

6. W. Rybczynski, *City Life: Urban Expectations in a New World* (New York: Simon & Schuster, 1995); J. Kunstler, *The Geography of Nowhere: The Rise and Decline of America's Man-made Landscape* (New York: Simon & Schuster, 1993); T. Hiss, *The Experience of Place* (New York: Random House, 1991).

7. Notwithstanding their insidious effect, some, like the noted American architect Robert Venturi, celebrate the commercial strips and sprawling development of places like Las Vegas as monuments to American populism and icons of their day.

8. See D. Hayden, *The Power of Place: Urban Landscapes as Public History* (Cambridge: MIT Press, 1995); T. Egan, "Urban Sprawl Strains Western States," *New York Times*, Dec. 29, 1997, A1.

9. S. Holmes, "Leaving the Suburbs for Rural Areas," *New York Times*, Oct. 19, 1997, A34; R. Collins, "Boomtown Moves to the Country," *Boston Globe*, Nov. 16, 1997, 1.

10. D. John, *Civic Environmentalism: Alternatives to Regulation in States and Communities* (Washington, D.C.: Congressional Quarterly Press, 1994), 289.

11. Progressive Policy Institute, *Executive Summary, Civic Environmentalism in Action: A Field Guide to Regional and Local Initiatives* (1999).

12. *See* W. Cronon, *Changes in the Land: Indians, Colonists, and the Ecology of New England* (New York: Hill and Wang, 1983); C. Merchant, *Ecological Revolutions: Nature, Science and Gender in New England* (Chapel Hill: University of North Carolina Press, 1989); T. Bender, *Toward an Urban Vision: Ideas and Institutions in Nineteenth Century America* (Baltimore: Johns Hopkins University Press, 1975).

13. D. Quammen, "Planet of Weeds: Tallying the Loss of Earth's Animals and Plants," *Harper's* (Oct. 1998): 68.

14. M. Dowie, *Losing Ground: American Environmentalism at the Close of the Twentieth Century* (Cambridge: MIT Press, 1995), 8.

15. R. Ford, "Behemoth on Wheels," *New York Times*, Aug. 20, 1997, A33.

Chapter 1

1. A. Leopold, *A Sand County Almanac* (New York: Oxford University Press, 1949).

2. J. Muir, *John Muir: In His Own Words* (Lafayette, Calif.: Great West Books, 1988).

3. N. MacLean, *A River Runs Through It and Other Stories* (Chicago: University of Chicago Press, 1976), 105.

4. D. Lerner and H. Lasswell, eds., *The Policy Sciences: Recent Developments in Scope and Method* (Stanford: Stanford University Press, 1951).

5. R. Williams, *Problems in Materialism and Culture* (London: Verso, 1980), 67.

6. W. Cronon, "Introduction: In Search of Nature," in W. Cronon, ed., *Uncommon Ground: Toward Reinventing Nature* (New York: Norton, 1995), 36.

7. W. Cronon, *Changes in the Land: Indians, Colonists, and the Ecology of New England* (New York: Hill and Wang, 1983), 13.

8. C. Merchant, *Ecological Revolutions: Nature, Gender and Science in New England* (Chapel Hill: University of North Carolina Press, 1989).

9. Ibid., 4–5, 19.

10. B. Staples, "The Master of Monticello," book review of Joseph J. Ellis's *American Sphinx: The Character of Thomas Jefferson*, *New York Times Book Review*, Mar. 23, 1997, 7.

11. T. Bender, *Toward an Urban Vision: Ideas and Institutions in Nineteenth Century America* (Baltimore: Johns Hopkins University Press, 1975), 4–5.

12. T. Jefferson to John Jay, Aug. 23, 1785, in J. P. Boyd, ed., *The Papers of Thomas Jefferson* (Princeton: Princeton University Press, 1952), 8:426.

13. Ibid., 4.

14. Tocqueville, *Democracy in America* (1835), ed. Phillips Bradley (New York: Alfred A. Knopf, 1945), 1:299–300.

15. Bender, *Toward an Urban Vision,* 24 (quoting T. Jefferson).

16. A. Runte, "The National Park Idea: Origins and Paradox of the American Experience," *Journal of Forest History* 21 (1977): 69.

17. H. Smith, *Virgin Land: The American West as Symbol and Myth* (Cambridge: Harvard University Press,1950), 253 (quoting F. J. Turner).

18. G. Wood, "The Bloodiest War," *New York Review of Books*, Apr. 9, 1998, 41.

19. F. J. Turner, *The Frontier in American History* (Washington, D.C.: Government Printing Office, 1894), 1.

20. W. Whitman, "Poem of the Sayers of the Words of the Earth," *Leaves of Grass,* quoted in Smith, *Virgin Land,* 44.

21. Ibid., ix.

22. W. Cronon, "The Trouble with Wilderness; Or, Getting Back to the Wrong Nature," in Cronon, *Uncommon Ground,* 77.

23. M. Sandel, *Democracy's Discontent: America in Search of a Public Philosophy* (Cambridge: Harvard University Press, 1996), 4.

24. B. Barber, *Strong Democracy: Participatory Politics for a New Age* (Berkeley: University of California Press, 1984), xiv.

25. L. Stout, *Bridging the Class Divide and Other Lessons for Grassroots Organizing* (Boston: Beacon Press, 1997).

26. C. Lasch, *The Revolt of the Elites and the Betrayal of Democracy* (New York: Norton, 1995), 8.

27. J. Dewey, *The Public and Its Problems* (1927; Athens, OH: Swallow Press, 1954), 148.

28. H. Putnam, "A Reconsideration of Deweyan Democracy," in M. Brint and W. Weaver, eds., *Pragmatism in Law and Society* (Boulder, CO: Westview Press, 1991), 125 (quoting J. Dewey and J. H. Tufts, *Ethics*).

29. R. Rorty, *Objectivity, Relativism and Truth: Philosophical Papers* (New York: Cambridge University Press, 1991), 1:196.

30. Sandel, *Democracy's Discontent,* 321.

31. Ibid., 349

32. A. Gopnik, "Olmsted's Trip," *New Yorker*, Mar. 3, 1997, 96.

33. Ibid.

34. R. Putnam, "Bowling Alone: America's Declining Social Capital," *Journal of Democracy* 6 (Jan. 1995): 67.

35. C. West, "The Limits of Neopragmatism," in *Pragmatism in Law and Society,* eds. M. Brint and W. Weaver (Boulder, Colo.: Westview Press, 1991), 125.

36. L. Whitt and J. Slack, "Communities, Environments, and Cultural Studies," *Cultural Studies* 8 (Jan. 1994): 21–22.

37. Barber, *Strong Democracy,* 306.

38. Putnam, "Bowling Alone," 76–77.

39. A. Kannapell, "Report Says Health of Society Lags Behind That of Economy," *New York Times*, Oct. 12, 1997, 34.

40. Putnam, "Bowling Alone," 65, 68.

41. Ibid., 70.

42. Ibid., 76.

43. R. Bellah, R. Madsen, W. Sullivan, A. Swidler, and S. Tipton, *Habits of the Heart* (New York: Harper & Row, 1985).

44. A. Wolfe, *One Nation, After All* (New York: Penguin, 1998).

45. Barber, *Strong Democracy,* xv.

46. Ibid., xiii.

47. C. Derber, W. Schwartz, and Y. Magrass, *Power in the Highest Degree: Professionals and the Rise of the New Mandarin Order* (New York: Oxford University Press, 1990), 209.

48. M. Lind, *The Next American Nation: The New Nationalism and the Fourth American Revolution* (New York: Free Press, 1995), 156.

49. J. Bonifaz and J. Raskin, *The Wealth Primary: Campaign Financing and the Constitution*, (Washington, D.C.: Center for Responsible Politics, 1994), xiii, 39.

50. U.S. Department of Commerce, Economics and Statistics Administration, Bureau of the Census, *1990 Census of Population, Social and Economic Characteristics* (Washington, D.C.: Government Printing Office, 1993).

51. W. Wilson, *When Work Disappears: The World of the New Urban Poor* (New York: Random House, 1996), 14.

52. O. Patterson, "Racism Is Not the Issue," *New York Times*, Nov. 16, 1997, 15.

53. P. Matthiessen, *In the Spirit of Crazy Horse* (New York: Viking Press, 1983), xxii.

54. Patterson, "Racism Is Not the Issue."

55. M. Oliver and T. Shapiro, *Black Wealth/White Wealth: A New Perspective on Racial Inequality* (New York: Routledge, 1995).

56. A. Harmon, "Racial Divide Found on Information Highway," *New York Times*, Apr. 16, 1998, A1, A22.

57. Wilson, *When Work Disappears*, xiii.

58. A. Hunt and A. Murray, "Rich Man, Poor Man," *Smart Money* Magazine (May 1998).

59. Lasch, *Revolt of the Elites*, 31.

60. Lind, *Next American Nation*, 154.

61. Derber, Schwartz, and Magrass, *Power in the Highest Degree*, 206.

62. Ibid., 6.

63. L. Menand, "The Trashing of Professionalism," *New York Times* Magazine, Mar. 5, 1995, 41.

64. Derber, Schwartz, and Magrass, *Power in the Highest Degree*, 4.

65. Menand, "Trashing," 41.

66. J. Heilemann, "Nerd File: The High-Tech High Life That Fuels the New Schmooze Industry," *New Yorker*, Apr. 20, 1998, 46.

67. Lasch, *Revolt of the Elites*, 6.

68. C. West, *Keeping Faith: Philosophy and Race in America* (New York: Routledge, 1993), 236.

69. L. Thurow, "One Nation, Divisible," *Boston Globe*, Sept. 9, 1997, C4.

70. C. Lindblom, *Politics and Markets* (New York: Basic Books, 1977), 356.

71. C. Reich, *Opposing the System* (New York: Crown Publishers, 1995), 49.

72. C. MacPherson, *The Political Theory of Possessive Individualism: Hobbes to Locke* (New York: Oxford University Press, 1963), 263–71.

73. J. Waldron, *The Right to Private Property* (New York: Oxford University Press, 1988), 356.

74. Thurow, "One Nation, Divisible."

75. Lind, *Next American Nation,* 190.

76. A. Wolfe, "Couch Potato Politics," *New York Times*, Mar. 15, 1998, 17.

77. Sandel, *Democracy's Discontent,* 332.

78. M. Pollan, "Living at the Office," *New York Times*, Mar. 14, 1997, A33.

79. J. K. Galbraith, *The Affluent Society* (Boston: Houghton Mifflin, 1958).

80. Lasch, *Revolt of the Elites,* 120.

Chapter 2

1. J. Ruskin, *Seven Lamps of Architecture* (London: Smith, Elder, 1849), 7.

2. P. Hawken, "Natural Capitalism," *Mother Jones* (Mar.–Apr. 1997).

3. D. Quammen, "Planet of Weeds: Tallying the Losses of Earth's Animals and Plants," *Harper's* (Oct. 1998).

4. W. Berry, *The Unsettling of America* (New York: Avon Books, 1977), 22.

5. L. Mumford, *The City in History: Its Origins, Its Transformations, and Its Prospects* (New York: Harcourt, Brace and Jovanovich, 1961).

6. D. Hayden, *The Power of Place: Urban Landscapes as Public History* (Cambridge: MIT Press, 1995), 43.

7. D. Abram, *The Spell of the Sensuous: Perception and Language in a More-Than-Human World* (New York: Random House, 1996), 266–71, 258.

8. E. V. Walter, *Placeways: A Theory of the Human Environment* (Chapel Hill: University of North Carolina Press, 1988), 204.

9. T. Hiss, *The Experience of Place: A New Way of Looking at and Dealing with Our Radically Changing Cities and Countryside* (New York: Random House, 1990), xi.

10. Hayden, *Power of Place,* 9.

11. R. Kaplan and S. Kaplan, *The Experience of Nature: A Psychological Perspective* (New York: Cambridge University Press, 1989).

12. T. Egan, "Drawing a Hard Line Against Urban Sprawl," *New York Times*, Dec. 30, 1996.

13. See M. Landy, M. Roberts, and S. Thomas, *The Environmental Protection Agency: Asking the Wrong Questions* (New York: Oxford University Press, 1994), 22; see also K. Jackson, *Crabgrass Frontier: The Suburbanization of America* (New York: Oxford University Press, 1985).

14. W. Cronon, *Changes in the Land: Indians, Colonists, and the Ecology of New England* (New York: Hill and Wang, 1983), 6.

15. T. Bender, *Toward an Urban Vision: Ideas and Institutions in Nineteenth Century America* (Baltimore: Johns Hopkins University Press, 1975), 29.

16. W. Rybczynski, *City Life: Urban Expectations in a New World* (New York: Simon & Schuster, 1995), 125.

17. Mumford, *City in History,* 447.

18. Rybczynski, *City Life,* 175.

19. J. Turner and J. Rylander, "Land Use: The Forgotten Agenda," in M. Chertow and D. Esty, eds., *Thinking Ecologically: The Next Generation of Environmental Policy* (New Haven: Yale University Press, 1997), 62.

20. J. Preston, "Battling Sprawl, States Buy Land for Open Space," *New York Times*, June 9, 1998, A23.

21. Turner and Rylander, "Land Use," 63.

22. T. Egan, "Drawing a Hard Line Against Urban Sprawl," *New York Times*, Dec. 30, 1996.

23. T. Egan, "Urban Sprawl Strains Western States," *New York Times*, Dec. 29, 1996.

24. B. McKibben, "Immigrants Aren't the Problem. We Are," *New York Times*, Mar. 9, 1998, A23.

25. S. Holmes, "Leaving the Suburbs for Rural Areas," *New York Times*, Oct. 19, 1997, 34.

26. Turner and Rylander, "Land Use," 60.

27. Rybczynski, *City Life,* 144.

28. A. Runte, "The National Park Idea: Origins and Paradox of the American Experience," *Journal of Forest History* 21 (1977).

29. Turner and Rylander, "Land Use," 63 (quoting A. Leopold).

30. Hawken, "Natural Capitalism," 43–45.

31. Ibid., 44.

32. U.S. EPA, *Information Packet About Toxic Release Inventory: Right to Know Brochure* (Washington, D.C.: Government Printing Office, 1996).

33. A. Aspelin, *Pesticide Industry Sales and Usage: 1992 and 1993 Market Estimates*, (Washington, D.C.: U.S. EPA, June 1994).

34. M. Dowie, *Losing Ground: American Environmentalism at the Close of the Twentieth Century* (Cambridge: MIT Press, 1995), 158.

35. C. Browner, speech to the All-States Meeting, Washington, D.C., June 4, 1996.

36. U.S. EPA, *State of the New England Environment* (Washington, D.C.: U.S. EPA, 1997), 7.

37. I. Kessel and J. O'Connor, *Getting the Lead Out: The Complete Resource on How to Prevent and Cope with Lead Poisoning* (New York: Plenum Trade, 1997), 7.

38. The term *brownfields* was coined by the Northeast-Midwest Institute, an independent, nonprofit regional policy center. See C. Bartsch et al., *New Life for Old Buildings: Confronting Environmental and Economic Issues to Industrial Reuse* (Washington, D.C.: Northeast-Midwest Institute, 1991).

39. A. Revkin, "For Urban Wastelands, Tomatoes and Other Life," *New York Times*, Mar. 3, 1998, A1, A21.

40. C. Bartsch and R. Munson, "Restoring Contaminated Industrial Sites," *Issues in Science and Technology* (Spring 1994): 74.

41. B. Goldman, *Not Just Prosperity: Achieving Sustainability with Environmental Justice* (Washington, D.C.: National Wildlife Federation, Feb. 1994), 11.

42. Ibid., 14.

43. "Farms Are Polluters of Nation's Waterways," *New York Times*, May 14, 1998, A19.

44. S. Burrington and B. Heart, *City Routes, City Rights: Building Livable Neighborhoods and Environmental Justice by Fixing Transportation* (Boston: Conservation Law Foundation, June 1998), 16.

45. C. Browner, press conference on Index of Watershed Indicators, Washington, D.C., Oct. 2, 1997.

46. L. Brown et al., *Vital Signs* (New York: Norton, 1996), 125

47. T. Dahl, *Wetland Losses in the United States, 1780's to 1980's* (Washington, D.C.: Fish and Wildlife Service, U.S. Department of the Interior, 1990).

48. T. Egan, "Where Water Is Power, the Balance Shifts," *New York Times,* Nov. 30, 1997, 1, 24.

49. J. Robbins, "Engineers Plan to Send a River Flowing Back to Nature," *New York Times*, May 12, 1998, C1.

50. T. Egan, "Urban Sprawl Strains Western States," *New York Times*, Dec. 29, 1996, 1.

51. J. Cushman, "Public Backs Tough Steps for a Treaty on Warming," *New York Times*, Nov. 28, 1997, 36.

52. Burrington and Heart, *City Routes,* 11.

53. Ibid., 10–11.

54. Brown et al., *Vital Signs,* 64.

55. K. Bradsher, "Light Trucks, Darlings of Drivers, Are Favored by the Law, Too," *New York Times*, Nov. 30, 1997, 1, 38–39.

56. U.S. Department of Transportation, Bureau of Transportation Statistics, *Transportation in the United States: A Review* (Washington, D.C.: U.S. Government Printing Office, 1997).

57. Egan, "Urban Sprawl," 1.

58. W. Cronon, "Introduction: In Search of Nature," in W. Cronon, ed., *Uncommon Ground: Toward Reinventing Nature* (New York: Norton, 1995), 25.

59. W. Cronon, "The Trouble with Wilderness; Or, Getting Back to the Wrong Nature," in ibid., 88.

60. Abram, *Spell of the Sensuous,* x.

61. Cronon, *Changes in the Land*, 159–60.

62. Hiss, *Experience of Place,* 206 (quoting R. Yaro).

63. R. Sennett, *The Corrosion of Character: The Personal Consequences of Work in the New Capitalism* (New York: Norton, 1998).

64. *See* J. Fritsch, "Stamford Wonders How to Turn a Mall Inside Out," *New York Times*, Sept. 22, 1997, A28.

65. Hayden, *Power of Place,* 99.

66. D. Kemmis, *Community and the Politics of Place* (Norman, OK: University of Oklahoma Press, 1990), 79.

67. A. Lupo, "Neighborhoods Works on Power Connection," *Boston Globe*, Sept. 14, 1997, D1.

68. T. Goldtooth, "Indigenous Nations: Sovereignty and Its Implications," in B. Bryant, ed., *Environmental Justice: Issues, Policies and Solutions* (Washington, D.C.: Island Press, 1995), 144.

69. R. Ford and R. Collins, "Boomtown Moves to the Country," *Boston Globe*, Nov. 16, 1997, 1; S. Holmes, "Leaving the Suburbs for Rural Areas," *New York Times*, Oct. 19, 1997, 34.

70. D. Massey and N. Denton, *American Apartheid: Segregation and the Making of the Underclass* (Cambridge: Harvard University Press, 1993), 14.

71. E. Blakely and M. Snyder, *Fortress America: Gated Communities in the United States* (Washington, D.C.: Brookings Institution Press, 1997).

72. B. Barber, "Big = Bad, Unless It Doesn't," *New York Times*, Apr. 15, 1998, A23.

73. Kemmis, *Community,* 137.

74. C. Cobb, T. Halstead, and J. Rowe, "If the GDP Is Up, Why Is America Down," *Atlantic Monthly,* Oct. 1995, 60.

75. Ibid., 65.

Chapter 3

1. A. Leopold, *A Sand County Almanac* (New York: Oxford University Press, 1968), 203, 224–25.

2. A British ambassador to the United States once called the national parks America's "best idea." Writer James Conaway adds that "the preservation of Federal real estate for the pleasure and benefit of the citizenry is, like freedom, a revolutionary impulse springing from the Enlightenment." J. Conaway, "Still Our Best Idea," *New York Times*, May 25, 1998.

3. L. Hughes, "Let America Be America Again," in A. Rampersad, ed., *The Collected Poems of Langston Hughes* (New York: Alfred A. Knopf, 1994).

4. M. Landy, M. Roberts, and S. Thomas, *The Environmental Protection Agency: Asking the Wrong Questions* (New York: Oxford University Press, 1994), 5.

5. G. P. Marsh, *Man and Nature: Or, Physical Geography as Modified by Human Action* (Cambridge: Harvard University Press, 1965), 35.

6. R. F. Nash, *Wilderness and the American Mind* (New Haven: Yale University Press, 1968), 92 (quoting Thoreau).

7. J. Muir, "The Tuolumne Yosemite in Danger," *Outlook* 87 (1907): 488.

8. J. Muir, "The National Park and Forest Reservations," *Harper's Weekly* 41 (1897): 567.

9. J. Muir, "The Treasure of Yosemite," *Century* 40 (1890): 488.

10. W. Cronon, "The Trouble with Wilderness; Or, Getting Back to the Wrong Nature," in W. Cronon, ed., *Uncommon Ground: Toward Reinventing Nature* (New York: Norton, 1995), 73, 76.

11. G. Pinchot, *The Fight for Conservation* (New York: Doubleday, Page, 1910), 45.

12. W. Shutkin, "The National Park Service Act Revisited," *Virginia Environmental Law Journal* 10 (1991): 354 (quoting Pinchot).

13. Pinchot, *Fight for Conservation,* 42, 50.

14. R. Gottlieb, *Forcing the Spring: The Transformation of the American Environmental Movement* (Washington, D.C.: Island Press, 1993), 26.

15. C. Jordan and D. Snow, "Diversification, Minorities, and the Mainstream Environmental Movement," in D. Snow, ed., *Voices from the Environmental Movement* (Washington, D.C.: Island Press, 1990), 76–77.

16. Gottlieb, *Forcing the Spring,* 30.

17. Ibid., 39–40.

18. K. Jackson, *Crabgrass Frontier: The Suburbanization of America* (New York: Oxford University Press, 1985).

19. M. Dowie, *Losing Ground: American Environmentalism at the Close of the Twentieth Century* (Cambridge: MIT Press, 1995), 23.

20. D. John, *Civic Environmentalism: Alternatives to Regulation in States and Communities* (Washington, D.C.: Congressional Quarterly Press, 1994), 5.

21. Landy, Roberts, and Thomas, *Environmental Protection Agency,* 3.

22. Ibid., 11.

23. W. Shutkin and C. Lord, "Environmental Law, Environmental Justice, and Democracy," *West Virginia Law Review* 96 (1994): 1117–18, n. 1.

24. E. Elliot, "Toward Ecological Law and Policy," in M. Chertow and D. Esty, eds., *Thinking Ecologically: The Next Generation of Environmental Policy* (New Haven: Yale University Press, 1997), 170.

25. M. Sandel, "It's Immoral to Buy the Right to Pollute," *New York Times*, Dec. 15, 1997, A29. See also Dowie, *Losing Ground,* 123.

26. Dowie, *Losing Ground,* 81 (quoting L. Billings).

27. M. Peterson, "Cleaning Up in the Dark," *New York Times,* May 14, 1998, D1.

28. U.S. Environmental Protection Agency, "Regulatory Reinvention (XL) Pilot Projects," *Federal Register*, Apr. 23, 1997, 19,872.

29. R. Russell, "Re-engineering Regulation," *Conservation Matters* 4 (Autumn 1997): 31.

30. Interview with Anne Kelly, Aug. 10, 1998.

31. Russell, "Re-engineering Regulation" 32 (quoting D. Struhs).

32. S. Lewis, "Feel Good Notions, Corporate Power and the Reinvention of Environmental Law," *Good Neighbor Project for Sustainable Industries,* Working Paper, Mar. 1997, 7–8.

33. Ibid., 9.

34. J. Cushman, Jr., "EPA and States Found to Be Lax on Pollution Law," *New York Times*, June 7, 1998, A1.

35. Russell, "Re-engineering Regulation," 32.

36. Ibid.

37. D. Esty and M. Chertow, "Thinking Ecologically: An Introduction," in *Thinking Ecologically*, 2.

38. Lewis, "Feeling Good Notions," 18

39. Gottlieb, "Forcing the Spring," 150–51.

40. Dowie, *Losing Ground,* 31.

41. Ibid., 124.

42. R. Bullard, "Overcoming Racism in Environmental Decisionmaking," *Environment* (May 1994): 11.

43. United Church of Christ Commission for Racial Justice and Public Data Access, *Toxic Wastes and Race in the United States: A National Report on the Racial and Socio-Economic Characteristics of Communities with Hazardous Waste Sites* (New York: United Church of Christ, 1987).

44. M. Lavelle and M. Coyle, "Unequal Protection: The Racial Divide in Environmental Law," *National Law Journal*, Sept. 21, 1993.

45. P. Shabecoff, "Environmental Groups Told They Are Racist in Hiring," *New York Times*, Feb. 1, 1990, A16.

46. Dowie, *Losing Ground,* 164 (quoting M. Fischer).

47. Interview with Vernice Miller, July 9, 1998.

48. Interview with Lois Adams, May 13, 1998.

49. J. Cushman, Jr., "Pollution Policy Is Unfair Burden, States Tell EPA," *New York Times*, May 10, 1998, 1, 16 (quoting W. Kovacs)

50. Dowie, *Losing Ground,* 126.

51. Cronon, "The Trouble with Wilderness," 85, 90 (emphasis in original).

52. W. Glaberson, "Novel Antipollution Tool Is Being Upset by Courts," *New York Times*, June 5, 1999, A1.

53. Ibid., A10 (quoting C. Pravlik).

54. L. Cole, "Empowerment as the Key to Environmental Protection: The Need for Environmental Poverty Law," *Ecology Law Quarterly* 19 (1992): 642, 643.

55. Ibid., 647.

56. Cronon, "The Trouble with Wilderness," *Uncommon Ground*, 81.

57. Dowie, *Losing Ground,* xiii.

58. C. West, *The American Evasion of Philosophy: A Genealogy of Pragmatism* (Madison: University of Wisconsin Press, 1989), 148–49.

59. C. Beitz, *Political Equality: An Essay in Democratic Theory* (Princeton: Princeton University Press, 1989), 97–119.

60. M. Briand, *Building Deliberative Communities,* (Charlottesville, Va: Pew Partnership for Civic Change, Spring 1995), 12–13.

61. H. Putnam, "A Reconsideration of Deweyan Democracy," in M. Brint and W. Weaver eds., *Pragmatism in Law and Society* (Boulder, CO: Westview Press, 1991), 232 (quoting J. Dewey).

62. A. Flint, "Leaders Fail to Guide Planning for Revival of Harbor, Some Say," *Boston Globe*, Apr. 17, 1998, B2.

63. Redefining Progress, Tyler Norris Associates, and Sustainable Seattle, *The Community Indicators Handbook: Measuring Progress Toward Healthy and Sustainable Communities*, (San Francisco: Redefining Progress, 1997), 2.

64. C. Powers and M. Chertow, "Industrial Ecology: Overcoming Policy Fragmentation," in *Thinking Ecologically*, 25.

65. W. McDonough and M. Braungart, "The Next Industrial Revolution," *Atlantic Monthly* (Oct. 1998).

66. W. McDonough, "The Hannover Principles," (University of Virginia, Charlottesville, Va.: UVA Architecture Publications, 1992).

67. W. Cudnohufsky and J. Abrams, *Summary Report: Londonderry Ecological Industrial Park*, Apr. 3, 1998.

68. W. Shutkin, "Environmental Justice and a Reconception of Democracy," *Virginia Environmental Law Journal* 14 (1995): 586 (quoting M. Delaney).

69. M. Sagoff, "Settling America: The Concept of Place in Environmental Politics," in P. Brick and R. Cawley eds., *A Wolf in the Garden: The Land Rights Movement and the New Environmental Debate* (Lanham, MD: Rowman & Littlefield, 1996), 254 (quoting A. Gussow).

70. Ibid., 255.

71. M. Dowie, "Environmentalism," *Estero* 2, no. 4 (1995) (quoting G. Snyder).

Chapter 4

1. S. Connor, *New England Natives: A Celebration of People and Trees* (Cambridge: Harvard University Press, 1994), 90–91.

2. S. Bass Warner, Jr., *Streetcar Suburbs: The Process of Growth in Boston (1870–1900)* (Cambridge: Harvard University Press, 1978), 17.

3. Ibid., 106–7.

4. Ibid., 101.

5. P. Medoff and H. Sklar, *Streets of Hope: The Fall and Rise of an Urban Neighborhood* (Boston: South End Press, 1994), 12–13.

6. Ibid.

7. Ibid., 16.

8. Ibid., 31.

9. Ibid., 32 (quoting a Boston Redevelopment Report).

10. Ibid., 33, 34 (quoting P. Bothwell).

11. Ibid., 86 (quoting W. Waldron).

12. G. Watson, *Urban Agriculture in Dudley Village: A Proposal to the Ford Foundation* (Boston: Dudley Street Neighborhood Initiative, May 1997), 3.

13. R. Foster, "Urbaculture: A New Word for an Old Business, Worldwide," *Boston Globe*, July 26, 1998, H29.

14. Interview with Greg Watson, Sept. 16, 1998.

15. M. Pollan, "Playing God in the Garden," *New York Times Magazine*, Oct. 25, 1998, 47.

16. Interview with Watson.

17. Interview with Trish Settles, Sept. 16, 1998.

18. Watson, *Urban Agriculture,* 2.

19. M. Porter, "The Competitive Advantage of the Inner City," *Harvard Business Review,* May 1995.

20. Interview with Watson.

21. G. Negri, "Fresh Produce Year-Round? It's Possible," *Boston Globe*, City Weekly Section, July 26, 1998, 1.

Chapter 5

1. G. Barth, *Instant Cities: Urbanization and the Rise of San Francisco and Denver* (New York: Oxford University Press, 1975).

2. A. Haupt, "Union Point Park: Waterfront Access, Recreation, and Equity," *Urban Ecologist*, no. 4 (1997): 10.

3. Bank of America, *Beyond Sprawl: New Patterns of Growth to Fit the New California* (Feb. 1995).

4. S. Burrington and B. Heart, *City Routes, City Rights: Building Livable Neighborhoods and Environmental Justice by Fixing Transportation* (Boston: Conservation Law Foundation, June 1998), 13.

5. Ibid.

6. Ibid., 14–15.

7. Ibid., 19.

8. J. Holtz Kay, *Asphalt Nation: How the Automobile Took Over America, and How We Can Take It Back* (New York: Crown Publishers, 1997), 40.

9. Ibid., 21.

10. Ibid., 46 (quoting C. Hayes).

11. Ibid., 47.

12. R. Bullard and G. Johnson, eds., *Just Transportation: Dismantling Race and Class Barriers to Mobility* (Philadelphia: New Society Publishers, 1997), xi (quoting J. Lewis).

13. Burrington and Heart, *City Routes,* 5.

14. Kay, *Asphalt Nation,* 305.

15. Much of the information regarding UC's history comes from my August 20, 1998, interview with Chris Hudson, project manager at the Fruitvale Development Corporation.

16. D. Kim, "Neighborhood Thinks Transit Villages Work," *Oakland Tribune* (quoting A. Martinez).

17. R. Knack, "BART's Village Vision," *Planning Practice* (Jan. 1995): 1 (quoting A. Martinez).

18. M. Bernick, "Can't Walk to Work? Then Walk to the Train," *Los Angeles Times*, May 4, 1993.

19. B. O'Brien, "This Time We Make the Plans," *Oakland Express*, Jan. 29, 1993, 2, 20 (quoting A. Martinez).

20. B. Wildavsky, "Transport Official Offers Oakland Help," *San Francisco Chronicle* (quoting M. Bernick).

21. Interview with Chris Hudson, Aug. 20, 1998.

22. Ibid.

23. Ibid.

24. Ibid.

25. K. Kirkwood, "Renewing the Waterfront, Better Days Ahead for Oakland Estuary," *Oakland Tribune*, Apr. 26, 1998, 1.

26. Interview with Michael Rios, Aug. 20, 1998.

27. Urban Ecology, *Blueprint for a Sustainable Bay Area* (San Francisco: Urban Ecology, Nov. 1996), 9.

28. See, e.g., R. Kaplan, *An Empire Wilderness: Travels into America's Future* (New York: Random House, 1998). Kaplan predicts a bleak future for the American West in which isolated suburbs, a disaffected populace, corporatist values, car culture, and racial enclaves carry the day, all amid a disfigured, degraded landscape. At first blush, his predictions ring true.

29. Urban Ecology, *Blueprint,* 5.

30. M. Dowie, "How It Might Have Been: Environmental Ad Hocracy in Coastal Marin," *Estero* 2, no. 4 (1995): 19 (quoting G. Snyder).

Chapter 6

1. W. Dimock, *Empire for Liberty: Melville and the Poetics of Individualism* (Princeton: Princeton University Press, 1989), 9 (quoting T. Jefferson).

2. R. Winks, *Frederick Billings: A Life* (New York: Oxford University Press, 1991), 92 (quoting G. Berkeley).

3. W. Stegner, *The Sound of Mountain Water* (New York: Doubleday, 1969), 38.

4. T. Egan, "Urban Sprawl Strains Western States," *New York Times*, Dec. 29, 1996, A1.

5. Sierra Club, *The Dark Side of the American Dream: The Costs and Consequences of Suburban Sprawl* (San Francisco: Sierra Club, Aug. 1998), 10. For broader treatments of the transformation of the West, see T. Egan, *Lasso the Wind: Away to the New West* (New York: Knopf, 1998) and R. Kaplan, *An Empire Wilderness: Travels into America's Future* (New York: Random House, 1998).

6. Sierra Club, *Dark Side,* 10.

7. T. Egan, "Drawing a Hard Line Against Urban Sprawl," *New York Times*, Dec. 30, 1996, A1.

8. Sierra Club, *Dark Side,* 10.

9. S. Holmes, "Leaving the Suburbs for Rural Areas," *New York Times*, Oct. 19, 1997, 34.

10. Ibid.

11. Egan, "Urban Sprawl," A1 (quoting R. Romer).

12. Sierra Club, *Dark Side,* 10.

13. M. Zeller, *Common Ground: Community-Based Conservation of Natural Resources* (May 1997), 4.

14. Stegner, *Sound of Mountain Water,* 38.

15. Conservation Partners, *Upper Elk River Valley: Protecting the Land and Sustaining Community* (Denver: Conservation Partners, Aug. 1994), 5.

16. Interview with Mike Tetreault, Oct. 7, 1998.

17. J. Brooke, "Rare Alliance in the Rockies Strives to Save Open Spaces," *New York Times*, Aug. 14, 1998, A18.

18. Interview with Susan Dorsey Otis, Oct. 7, 1998.

19. F. Williams, "Planning for Space: Saving Farmland and Building Community in Colorado's Yampa Valley," *Chronicle of Community* (Winter 1998): 7 (quoting A. Stettner).

20. Ibid.

21. Conservation Partners, *Routt County Open Lands Plan*, (Denver: Conservation Partners, June 1995,) 8.

22. Ibid.

23. *See* J. Brooke, "It's Cowboys vs. Radical Environmentalists in the New Wild West," *New York Times*, Sept. 20, 1998, 31.

24. Interview with Lynne Sherrod, Oct. 7, 1998.

25. Ibid.

26. H. Clifford, "Conservation Groups Ropes in a Working Ranch," *High Country News*, Nov. 27, 1995, 12 (quoting J. Williams).

27. Interview with Mark Burget, Oct. 6, 1998.

28. G. Low, *A Colorado Citizen's Guide to Achieving a Healthy Community, Economy and Environment* (Leesburg, Va.: The Nature Conservancy, 1996), 1.1.

29. Interview with Jay Fetcher, Oct. 7, 1998.

30. Interview with Marty Zeller, Oct. 6, 1998.

31. Conservation Partners, *Upper Elk River Valley*, 8.

32. Williams, "Planning for Space," 13.

33. Ibid., 7.

34. Ibid., 12 (quoting A. Baur and Zeller).

35. C. Cornelius, "Beef Ranchers Take Different Routt," *Denver Post*, June 22, 1998, B1.

36. Ibid. (quoting C. J. Mucklow).

37. D. Kemmis, *Community and the Politics of Place* (Norman, OK: University of Oklahoma Press, 1990), 6–7.

38. Egan, *Lasso the Wind*, 247.

Chapter 7

1. M. Courtney, "State Planners Seek to Defuse Opposition," *New York Times*, Apr. 3, 1988, 1 (quoting J. Gilbert).

2. J. Kunstler, *Home from Nowhere: Remaking Our Everyday World for the 21st Century* (New York: Simon & Schuster, 1996), 257.

3. J. Kunstler, *The Geography of Nowhere: The Rise and Decline of America's Man-Made Landscape* (New York: Simon & Schuster, 1993), 47–48.

4. Ibid., 48.

5. R. Moe and C. Wilkie, *Changing Places: Rebuilding Community in the Age of Sprawl* (New York: Henry Holt, 1997), 46.

6. Ibid., 50 (quoting L. Mumford).

7. B. Lawrence, "New Jersey Needs to Take Steps to Liberate Itself from its 'Car Culture,'" *Trenton Courier News*, Sept. 27, 1998. California and many other western states do not have toll roads.

8. New Jersey Future, *Living with the Future in Mind: The Sustainable State 1995 Program Report* (Trenton, N.J.: New Jersey Future, May 1996), 11.

9. Ibid., 10–11, 33, 35.

10. G. Hardin, "The Tragedy of the Commons," *Science*, 162 (1968).

11. R. Hanley, "To Preserve Open Space, More New Jerseyans Are Supporting New Local Taxes," *New York Times*, May 20, 1998, A25.

12. Ibid. (quoting A. Calotta and M. Kryger).

13. Interview with Tom Czerniecki, Nov. 12, 1998.

14. Interview with Barbara Davis, Nov. 17, 1998.

15. Interview with Carol Rufener, Nov. 16, 1998.

16. Interview with Denise Coyle, Nov. 10, 1998.

17. D. Hilgen, "'Work Together' Is Theme from Somerset Summit," *Somerset County Courier-News*, Apr. 17, 1997 (quoting R. Bateman).

18. M. McGarrity, "Growth Stages: The Cross-acceptance Process Aims for Consensus on Development," *Newark Star-Ledger*, Mar. 15, 1998.

19. Somerset County Economic Development Summit, *Report to the Community* (Somerville, N.J.: Somerset County Board of Chosen Freeholders, 1997), 1.

20. Ibid., 3.

21. Somerset County Economic Development Summit II, *Report to the Community* (Somerville, N.J.: Somerset County Board of Chosen Freeholders, June 1998).

22. Somerset County Chamber of Commerce, *Somerset Coalition for Smart Growth Business Plan* (Somerville, N.J.: Somerset County Board of Chosen Freeholders, 1998).

23. Interview with Coyle.

24. Interview with Barbara Lawrence, Dec. 2, 1998.

25. New Jersey Future, *Draft Goals and Indicators for a Sustainable State* (Trenton, N.J.: New Jersey Future, Nov. 1998), 1.

26. New Jersey Future, *Background Paper: Leadership Conference on Indicators for Sustainable Development in New Jersey* (Trenton, N.J.: New Jersey Future, 1997), 3.

27. Ibid., 5.

28. Ibid.

29. New Jersey Future, *Living with the Future in Mind: Goals and Indicators for New Jersey's Quality of Life, 1999 Sustainable State Project Report* (Trenton, N.J., New Jersey Future, May 1999).

30. New Jersey Future, *Background Paper*, 11.

31. Ibid.

32. Ibid.

33. New Jersey Future, *Living with the Future in Mind*, 20 (quoting R. Currie).

34. Ibid. (quoting V. Miller).

35. Ibid., 23–24 (quoting V. Miller).

36. J. Kunstler, *The Geography of Nowhere: The Rise and Decline of America's Man-Made Landscape* (New York: Simon and Schuster, 1993).

37. M. Orfield and D. Rusk, *Metropolitics: A Regional Agenda for Community and Stability* (Washington, D.C.: Brookings Institution Press, 1997).

38. L. Brown, *Vital Signs 1996* (New York: Norton, 1996), 99.

39. W. Stevens, "One in Every 8 Plant Species Is Imperiled, a Survey Finds," *New York Times*, Apr. 9, 1998, A1, A24.

40. D. Meadows, J. Randers, and W. Behrens III, *The Limits to Growth* (New York: Universe, 1972); D. Meadows, D. Meadows, and J. Randers, (Post Mills, VT: Chelsea Green, 1992); B. McKibben, *The End of Nature* (New York: Random House, 1989), *Hope, Human and Wild: True Stories of Living Lightly on the Earth* (Minneapolis: Hungry Mind Press, 1997), and *Maybe One: A Personal and Environmental Argument for Single-Child Families* (New York: Simon & Schuster, 1998).

41. R. Braile, "Bulldozing Our Way to a Greener World," *Boston Globe*, Apr. 18, 1999, D5 (quoting D. Meadows).

42. Ibid.

Chapter 8

1. J. Cassidy, "The Woman in the Bubble," *New Yorker*, Apr. 26, May 3, 1999, 56, 58.

2. A. Spirn, "Constructing Nature," in W. Cronon, ed., *Uncommon Ground*, (New York: Norton, 1995), 110 (quoting Marsh).

3. L. Hughes, "Let America Be America Again," in A. Rampersad, ed., *The Collected Poems of Langston Hughes* (New York: Alfred A. Knopf, 1994), 190.

Index

Abram, David, xiii, 48, 73
ACE (Alternatives for Community & Environment), 4, 5–8
Adams, Lois, 118–119
Addams, Jane, 97, 115
African-Americans
  air pollution and, 69–70, 170
  in Dudley Area of Boston, Massachusetts, 145–146
  industrial hazards and, 65
  poverty and, 35–37
  Warren County, North Carolina, 113
Agrarian republicanism, 24
Ailanthus altissimus (tree of heaven), 80
Air pollution, 48
  ambient, 68–71
  automobiles and, 70–71
  in California, 68–69, 170, 172–173
  consumption and, 58–59
  control, 59–60
  environmental indicators of, 224
  fine-particulate matter and, 122
  impact, 130
  indoor, 71–72
  South Bay area, Boston, 7–8
  transportation and, 56
Alternatives for Community & Environment (ACE), 4, 5–8
Architecture, civic, 46
Aristotle, 31
Asbestos, 61
Automobiles, air pollution and, 70–71
Banana Kelly Community Improvement Association, 117, 118
Barber, Benjamin, 28–29, 31, 34, 84
Bay Area, California. *See* San Francisco, California
Bay Area Rapid Transit authority (BART), 175–182
Beck, Ulrich, 89
Bellah, Robert, 33–34
Bender, Thomas, 24
Berkeley, California, 190
Berkeley, George, 190
Bernick, Michael, 177
Berry, Wendell, xiii, 47, 143
Billings, Leon, 103–104
Blakely, Edward, 83
Bonifaz, John, 35
Boston, Massachusetts
  Board of Health, 8–9
  Dudley area. *See* Dudley Area
  historical aspects, 143–149
  public market, 164–165
  South Bay area, 7
  urban development in, 9–11
Bothwell, Paul, 147
Briand, Michael, 129
Bronx Community Paper Company, 117
Brook Avenue greenhouse-bioshelter, 157, 159, 161
Brower, David, 67
Browner, Carole, 64–65

Brownfields, 11, 50
- adverse effects of, 64–65
- definition of, 63
- industrial ecology and, 139
- location of, 64–65
- locations of, 104
- number of, 64
- public health risk of, 64
- redevelopment, 119, 159–161
- toxic materials and, 60–65

Bryant, Hayley, 167
Bullard, Robert, 80, 115, 119
Burget, Mark, 198
Burrington, Steve, 173

CAAP (Coalition Against the Asphalt Plant), 8, 9
California
- Bay Area, 170–172 *See also* Oakland, California; San Francisco, California
- Environmental Quality Act, 173
- Los Angeles, 68–69, 113
- transit-oriented development, 176–177
- transportation policy, 173–174

Calotta, Anita, 214
Carbon emissions, 70
Carson, Rachel, 61, 98
Catamount proposal, 196
CCLT (Colorado Cattlemen's Land Trust), 202
Center for Health, Environment and Justice, 112
Charles River, Massachusetts, 68
Chertow, Marian, 108
Citizen's Clearinghouse for Hazardous Waste, 112
Citizen suits, 121–122
City Beautiful movement, 55
"Civic alchemy," 162
Civic democracy
- description of, 28–31
- overall state of, 32
- political participation and, 34–35
- privatization and, 41–44
- public investment and, 41–44
- racial equality and, 35–37
- social capital and, 32–34
- social condition of, 45–46
- socioeconomic equality and, 37–41

Civic engagement, 228–229
Civic environmentalism, xiv
- challenges to, 240–243
- community/regional planning and, 131–134
- core concepts, 128–141, 240
- definition of, xiv
- description of, 13–20
- education and, 135–136
- as emergent paradigm, 140–141
- environmental justice and, 139–140
- environmental organizations and, 239
- focus of, 238
- Fruitvale Transit village and, 184–187
- fundamentals of, 18–19
- historical/ideological roots, 21–26
- industrial ecology and, 136–139
- momentum from, 243–244
- New Jersey's sustainability efforts and, 230–235
- participatory process and, 128–131
- place and, 140
- privatization and, 2, 242
- Routt County conservation-based development and, 206–208
- strategies, forms of, 238–239
- systems approach of, 19, 22, 133
- terminology, 15
- urban agriculture and, 162–165

Civic health
- core indicators, 32–44
- decline, environmental effects of, 86–87
- environmental quality and, 45–49

Civic organizations. *See also specific civic organizations*
- secondary, 33
- tertiary, 33
- third-sector, 33

Civil Rights Act, 119
Civil rights laws, 172
Clean Air Act, 101–102, 103, 108
Clean Water Act, 102, 108

CLF (Conservation Law Foundation), 239
Coalition Against the Asphalt Plant (CAAP), 8, 9
Cole, Luke, 123–124
Colorado
 Douglas County, growth in, 191
 environmental regulations, 193–194
 environmental/social changes in, 192–193
 The Nature Conservancy, 198–199
 Responsible Growth Act, 193
 Routt County. *See* Routt County, Colorado
 urbanization in, 191
Colorado Cattlemen's Land Trust (CCLT), 202
Command-and-control regulation, 101, 103, 104
"Common place civilization," 11, 30
Communities
 democratic sustainable, 11, 30
 disempowerment of, 77
 environmental action. *See* Civic environmentalism
 environment and, 31
 ideal, 5
Community indicators, 133–134
Community Indicators Network, 133
Community institutions, decline in, 37
Community planning
 civic environmentalism and, 131–134, 240
 feedback mechanism for, 133–134
Community-supported agriculture (CSA), 158
Comprehensive Environmental Response, Compensation and Liability Act (Superfund), 104–105, 150
Concerned Citizens in Action, 113
Conservation, 93–94
Conservation Law Foundation (CLF), 239
Consumption
 development and, 50–51
 environmental impact of, 58–59
 production and, 58–59
Corporations
 economic growth and, 85–86
 political participation of, 78, 79
Coyle, Denise, 221
Cronon, William, 23, 24, 27, 72–73, 92, 120–121, 127, 237
Cross-acceptance, 219
CSA (community-supported agriculture), 158
Culture
 environment and, 23
 nature and, xvi
Czerniecki, Thomas, 214, 215

Davis, Barbara, 215
DDT, 61, 63
Deer, protection of, 213–214
Delaney, Michael, 139–140
Democracy
 American, environmentalism and, 89–91
 civic. *See* Civic democracy
 civic environmentalism and, 240
 environmentalism and, 120–126
 environmental theory of, 28–31
 environment and, 1, xvi
 pursuing, 237–240
 social/environmental fate of, 1–2
 strong, 28–29, 129
Derber, Charles, 39
Development
 conservation-based, 13
 consumption and, 50–51
 definition of, 50
 efficient, as environmental indicator, 223
 environmental change and, 50–55
 limited, 200–201
 place-based, 13
 private property, liberalism and, 54–55
 production and, 50–51
 reserved, 200
 transit-based, 13
 in United States, history of, 51–54

DeVillars, John, 106
Dewey, John, 21, 29, 89, 129–130
Digital technology, 43–44
Dioxin, 61
Dowie, Mark, 89, 112, 120, 127–128
Downing, Andrew Jackson, 52
DSNI. *See* Dudley Street Neighborhood Initiative
Dudley Area, Boston, Massachusetts
  Brook Avenue greenhouse-bioshelter, 157, 159, 161
  DSNI. *See* Dudley Street Neighborhood Initiative
  family income in, 152
  historical aspects, 143–149
  pollution problems in, 152
  PRIDE, 152
  urban renewal programs, 146–148
Dudley Street Neighborhood Initiative (DSNI), 19
  brownfield redevelopment, 159–161
  Don't Dump on Us campaign, 148
  environmental legislation and, 150–152
  formation, 148
  lead poisoning and, 152–153
  Massachusetts Highway Department and, 153–154
  pollution problems and, 149–150
  revitalization plan, 148–149
  Supplemental Environmental Project, 153–154
  urban agricultural projects, 153–154, 155
  Winthrop Estates, 149

Earth Day, 98
Easterbrook, Gregg, 102
Ecoindustrial parks (EIPs), 138
"Ecological revolutions," 23–24
Ecology, community and, 89
Economic growth, corporate power and, 85–86
Ecosystems, environmental indicators of, 223
Education
  civic environmentalism and, 240
  environmental, 135–136
Egan, Timothy, 49, 68, 189
EIPs (ecoindustrial parks), 138
Elliot, E. Donald, 101, 103
Emergency Planning and Community Right-to-Know Act (EPCRA), 62
Endangered species, 109
Endangered Species Act, 194
Environment. *See also* Environmental change
  American, ideology of, 24–26
  community and, 31
  culture and, 23
  democracy and, 1, xvi
  Native American's relationship with, 73
  preservation of, 92–93
  quality, civic health and, 45–49
  self-concept and, xiv–xv
Environmental blackmail, 80, 115
Environmental change
  civic causes of, 75–86
  civic health and, 86–87
  consumption and, 58–59
  from development, 86
  development and, 50–55
  environment quality and, 72–75
  political participation and, 78–79
  privatization and, 83–86
  production and, 55–58
  public investment and, 83–86
  racial equality and, 80–81
  social capital and, 76–77
  social/economic forces in, 3, 23–24
  socioeconomic equality and, 82–83
  statistics on, 126–127
  as threat to American society, 90–91
Environmental education, 135–136
Environmental impacts, 59–60
Environmentalism, xv
  as American democracy symbol, 89–91
  civic. *See* Civic environmentalism
  civic issues and, 26–28
  democracy and, 120–126
  emerging model of, 4–13
  grass-roots, 96–97, 111–114
  holistic approach, 22

mainstream-professional, 97–99, 111–114, 116–117, 122–123
new approaches for, 126–128
pursuing, 237–240
Romantic-Progressive, 91–97
traditional, failure of, 6
in twenty-first century, xiii–xiv
Environmental justice
civic environmentalism and, 139–140, 240
critique of, 114–120
Environmental law and policy system
description of, 101–105
EPA and, 99–101
Environmental organizations. *See also specific environmental organizations*
civic environmentalism and, 239
Environmental protection, lingering gaps in, 108–110
Environmental Protection Agency (EPA)
definition of brownfields, 63
emissions standards, 70
environmental law and policy system, 99–101
environment justice programs, 118–119
fine-particulate matter regulation, 122
hazardous chemicals and, 62
history of, 101
mainstream-professional environmentalism and, 116
mission of, 100
Office of Civil Rights, 119
Project XL, 105–106
Superfund and, 104–105
Environmental Quality Act, California, 173
Environmental regulations, 57
Colorado, 193–194
command-and-control, 101, 103, 104
DSNI and, 150–152
for pollution control, 100
reinvented approaches for, 105–108
Environmental Results Program (ERP), 106–107, 108
Environment quality, environmental change and, 72–75
EPA. *See* Environmental Protection Agency
EPCRA (Emergency Planning and Community Right-to-Know Act), 62
ERP (Environmental Results Program), 106–107, 108
Esty, Daniel, 108
Everdell, Ros, 148

Fetcher, Jay, 199, 202, 207
Fischer, Michael, 116
Floods, 68
Food Project, Roxbury, Massachusetts, 153
Fordham University Institute for Innovation in Social Policy, 32
Fruitvale Community Collaborative, 174–175
Fruitvale Recreation and Open Space Initiative (FROSI), 182–183
Fruitvale Transit village, civic environmentalism and, 184–187

Galbraith, John Kenneth, 44
Garfield, James R., 94
GDP (gross domestic product), 85–86
General Agreement on Tariffs and Trade, 39–40
Gibbs, Lois, 112, 113
Global capitalism
civic environmentalism and, 240–241
mainstream-professional environmentalism and, 125
socioeconomic equality and, 38, 39–40
Gottlieb, Robert, 95, 96
Government
decentralization of, 2
traditional role of, 2
Grass-roots environmentalism, 96–97, 111–114
Greeley, Horace, 25
Green design, 137
Greenhouse gases, 70
Gross domestic product (GDP), 85–86
Group of Ten, 114, 116

Hamilton, Alice, 97

Hannover Principles, 137–138
Hanson, Fred, 108
Harris, Elihu, 177
Haskell, Llewellyn, 210
Hawken, Paul, 46, 58–59
Hayden, Dolores, 47, 48, 49, 77, xiv
Health, environmental exposures and, 63
Heart, Bennet, 173
Hegel, G. W., 42
Herschkowitz, Allen, 117
Hetch Hetchy Valley, Yosemite National Park, 94
Hill, Tennessee Valley Authority v., 111–112
Hiss, Tony, 48
Hoffman, Donna, 36–37
Housing development, affordable, 215–216
Hudson, Chris, 179, 182, 185
Hughes, Langston, xvi, 14, 90, 243–244

IE. *See* Industrial ecology
Income inequality, 38
Industrial ecology (IE)
  civic environmentalism and, 136–139, 240
  definition of, 11–12
  Urban Agricultural Strategy and, 163
Industrialization, 48, 52, 62–63
Intel Corporation, 107
Internet, 36–37, 241
Intersurface Transportation Equity Act (ISTEA), 177

Jablonski, David, 17
Jacobs, Jane, 11, 30, 31, 77
Jefferson, Thomas, 24–25, 31, 90, xvi
John, DeWitt, 15
Johnson, Mark, xvi
Jordan, Barbara, xv
Justice, environmental, 139–140

Kaplan, Rachel, 49
Kaplan, Stephen, 49
Kay, Jane Holtz, 174
Kelly, Anne, 105–106
Kemmis, Daniel, 77, xv
King, Martin Luther, Jr., 1
Kovacs, William, 119
Kryger, Melinda, 214
Kunstler, James, 210–211, 231

Laissez-faire era, 42
Lakoff, George, xvi
Land, xv–xvi
"Land grabs," 241
Land stewardship, xvi
Land use. *See* Development
Landy, Marc, 90, 100–101
Lasch, Christopher, 29, 39, 44
Latinos, industrial hazards and, 65
Law, civic environmentalism and, 242–243
Lawrence, Barbara, 222, 223, 227–228
Lead poisoning, 61, 152–153
Leopold, Aldo, 22, 55, 241
Lewis, John, 173
Lewis, Sanford, 107, 109
Liberalism, 41–42, 54–55
Limited development, 200–201
Lind, Michael, 35, 39
Locke, John, 42
Los Angeles, California, 68–69, 113
Louisiana Purchase of 1803, 25
Love Canal, New York, 112
Low, Greg, 199
Low-income communities, industrial hazards and, 65

McDonough, William, 137
MacKaye, Benton, 97
McKibben, Bill, 53, 234
MacLean, Norman, 22
Magrass, Yale, 39
Mainstream-professional environmentalism
  failure of, 122–125
  global capitalism and, 125
  legal/technical nature of, 122–123
  privatization and, 125
  public investment and, 125–126
Manchester, New Hampshire, 10–11
Marsh, George Perkins, 1, 91

Marshall, Robert, 95–96
Martinez, Arabella, 174, 176–177
Massachusetts
  Boston. *See* Boston, Massachusetts
  Department of Environmental Protection, 106–107
  Environmental Policy Act, 151
  Highway Department, DSNI and, 153–154
Mass transit, 171–172
Meadows, Donella, 234
Menand, Louis, 38–39
Merchant, Carolyn, 23–24, 47
Metaphors, xvi
Miller, Vernice, 117, 118, 229
Mining, 56
Minorities. *See also* African-Americans
  air pollution and, 69–70
  in Dudley Area of Boston, Massachusetts, 145–146
  environmental hazards and, 80–81
  Fruitvale Transit village and, 185–186
  industrial hazards and, 65
  poverty and, 35–37
  segregation and, 2–3
Money, in politics, 35
Morresi, Angelo, 209
Morris County, New Jersey
  Green Acres program, 214
  open space initiatives, 213–218
  watershed protection initiatives, 216–217
Mucklow, C.J., 197, 202, 204–205, 207
Muddy River, Massachusetts, 68
Muir, John, 22, 92, 94–95
Mumford, Lewis, 47–48, 52, 211, 237

NAAQS (National Ambient Air Quality Standards), 103–104, 121
Nartional Wildlife Federation, 111
Nathan Cummings Foundation, 153
National Ambient Air Quality Standards (NAAQS), 103–104, 121
National Audubon Society, 111, 112, 116, 239
National park system, 55, 90
National Wildlife Federation, 112
Native Americans, 80, 139
Natural capital, 46–47
Natural resources, environmental indicators of, 223–224
Nature, culture and, xvi
The Nature Conservancy (TNC), 198–199
NEI (New Ecology, Inc.), 13
Neighborhoods United for South Bay (NUSB), 10, 163
New Ecology, Inc. (NEI), 13
New Jersey
  agricultural history, 209–210
  cross-acceptance process, 219–220
  environmental indicators, 223–225
  highway system, 212
  historical aspects, 210–211
  Morris County. *See* Morris County, New Jersey
  Randolph Township, 214–217
  Somerset County, 218–222
  suburbanization/environmental change, 211–213
  sustainability, challenges of, 226–227
  Sustainable State Project, 222–230
  sustainablity efforts, civic environmentalism and, 230–235
Newman, Penny, 113
New York City, South Bronx, 117–118
North American Free Trade Agreement, 39–40
NRDC (Natural Resources Defense Council), 117–118
NUSB (Neighborhoods United for the South Bay), 10, 163

Oakland, California
  air pollution in, 170
  environmental protection, civil rights laws and, 172
  historical aspects, 168–169
  mass transit in, 171–172
  revitalization efforts, goals/principles of, 180–181
  transportation policy in, 169–170
  waterfront access, restoring, 182–184
Occupational exposure, to toxics, 63

Oldenburg, Ray, 30
Oliver, Melvin, 36
Olmsted, Frederick Law, 11, 26, 30, 54–55, 80, 83
O'Neil, Tip, 31
Open space
 acquisition, 233–234
 initiatives, 213–218
 terminology, 233
Orfield, Myron, 229–230
Otis, Susan Dorsey, 196, 197, 202, 207
Overdevelopment, 54–55

Patterson, Orlando, 36, 45
PCBs (polychlorinated biphenyls), 61, 63, 113
Peña, Federico, 178
Pesticides, 62
Petroleum, 61
Pinchot, Gifford, 93–94
Place
 civic environmentalism and, 140, 240
 Fruitvale Transit village and, 186
 power of, 49
 sense of, 48, 49
 structure of, 48
 Urban Agricultural Strategy and, 163
Place-bound identities, 48
Point sources, 102
Political participation, 34–35, 78–79
Politics
 liberal, 41–42
 local, 31
 money in, 35
Pollan, Michael, 43, 155–156
Pollution. *See* Air pollution; Water pollution
Pollution control methods, 100, 137
Polychlorinated biphenyls (PCBs), 61, 63, 113
Population
 growth, 52
 sprawl, 53–54, 169, 191–192
Porter, Michael, 158
Poverty, 35–37, 38
Power of place, 49
PPI (Progressive Policy Institute), 16
Preservation, environmental, 92–93, 94
Preservationists, 92–93. *See also specific preservationists*
Private property liberalism, development and, 54–55
Privatization
 civic democracy and, 41–44
 civic environmentalism and, 2, 242
 environmental change and, 83–86
 mainstream-professional environmentalism and, 125
Production
 consumption and, 58–59
 definition of, 55
 development and, 50–51
 environmental impact of, 55–58
 life cycle approach, 137
Professionalism, 38–39
Progressive Policy Institute (PPI), 16
Project XL, 105–106, 107
Public consciousness, 125–126
Public discourse, decline of, 242
Public health risk, of brownfields, 64
Public interest organizations, 110–111, 112. *See also specific public interest organizations*
Public investment
 civic democracy and, 41–44
 environmental change and, 83–86
 mainstream-professional environmentalism and, 125–126
Putnam, Robert, 30, 31, 32–33

Qaummen, David, 47

Race
 equality, 35–37, 80–81
 inequality, 2–3, 80–81, 115–116
Redefining Progress, 133
Regionalism, 229
Regional planning, civic environmentalism and, 131–134
Reich, Charles, 40
Reserved development, 200

Rios, Michael, 183, 185, 186
Roberts, Marc, 90, 100–101
Rocky Mountain West. *See also* Routt County, Colorado
  historical aspects, 189–190
  urbanization, 190–192
Roosevelt, President Theodore, 94
Rorty, Richard, 29
Rousseau, Jean-Jacques, 93
Routt County, Colorado, 191–192
  Catamount proposal, 196
  community preservation strategy, 195–201
  conservation-based development, civic environmentalism and, 206–208
  Open Lands Plan, 203–205
  Open Lands Steering Committee, 202–203
  rancher/environmentalist ties in, 197–199
  rural nature of, 194–195
  seasonal population, 195
  Upper Elk River Valley Compact, 201–202
Roxbury, Massachusetts, 10
  as agricultural center, 164–165
  Food Project, 153
  historical aspects, 143–146
  land use-related problems, 151
Ruckelshaus, William, 101
Ruefener, Carol, 217
Ruskin, John, 46
Russell, Rusty, 108
Rybczynski, Witold, 52

Sagoff, Mark, 140
Sandel, Michael, 28, 29–30
San Francisco, California
  air pollution in, 170
  Bay Area Rapid Transit authority, 175–182
  historical aspects, 167–169
  motor vehicle accidents and, 170–171
  sprawl, 169
  transportation infrastructure, 171–172
Schama, Simon, xiv–xv
Schwartz, William, 39
SCSG (Somerset Coalition for Smart Growth), 221
Segregation, 2–3, 80–81
Settles, Trish, 152, 154, 156, 163
Shapiro, Thomas, 36
Sherrod, Lynne, 189, 197–198, 199, 202, 207
Sierra Club, 95–96, 110–111, 112, 116, 239
Silent Spring (Carson), 61, 98
Slack, Jennifer Daryl, 31
Smart growth, 11, 234–235
Snyder, Gary, 140
Snyder, Mary Gail, 83
Social capital
  civic democracy and, 31, 32–34
  decline in, 33–34, 45–46
  definition of, 30
  environmental decline and, 76–77
  natural capital and, 46–47
Social networks, decline in, 37
Society, third places of, 30
Socioeconomic equality
  civic democracy and, 37–41
  environmental decline and, 82–83
Somerset Coalition for Smart Growth (SCSG), 221
Somerset County, New Jersey, 218–222
  cross-acceptance process and, 219–220
  first economic summit, 219–220
  overdevelopment, 218
  second economic summit, 221
  smart-growth strategy, 221–222
South Bay Area, Boston, Massachusetts, 163
  air pollution problem in, 7–8
  incinerator site, 9–10
South Bronx Clean Air Coalition, 118
Special interest groups, 78
Sprawl, 53–54, 169, 191–192
SSP. *See* Sustainable State Project
Stakeholders, 14
Staples, Brent, 24
Steamboat Springs, Colorado, 195
Stegner, Wallace, 194

Stettner, Arianthe, 196, 206
Stonyfield Farm Yogurt Company, 10–11, 138
Stout, Linda, 29
Stride Rite Shoe Company, 10
Studer, Ray, 192
Suburbanization, 50, 52, 53–54
Superfund (Comprehensive Environmental Response, Compensation and Liability Act), 104–105, 150
Superfund sites. *See* Brownfields
Supplemental Environmental Project (SEP), 153–154
Sustainable design, 137–138
Sustainable State Project (SSP), 222–231
Systems theory, 22

Technology, 48
Tennessee Valley Authority v. Hill, 111–112
Tennyson, Lord, 45
Tetreault, Mike, 199
Thomas, Stephen, 90, 100–101
Thoreau, Henry David, 91
Thurow, Lester, 40
Title VI, 119
TNC (The Nature Conservancy), 198–199
Tocqueville, Alexis de, 25, 31, 34, 49, 90
Todd, John, 156–157
Todesca Equipment Company, 7–8
Toxic materials
  brownfields and, 60–65
  water pollution and, 65–68
  wetlands degradation and, 65–68
Toxic Release Inventory, 28, 134
Toxics Use Reduction Act, 134
Toxic Wastes and Race in the United States, 115–116
Transit-oriented development, 176–177
Transportation
  in California, 169–170, 173–174
  efficient, as environmental indicator, 223
  pollution and, 56
  in production process, 56
Tree of heaven *(Ailanthus altissimus),* 80
Truckee irrigation project, 67
Turkel, George, 156–157
Turner, Frederick Jackson, 25–26

UAS. *See* Urban Agricultural Strategy
Unemployment rates, 37–38
United States. *See also specific states*
  environmental movement in, 91
  history of development in, 51–54
  land ethics in, 75
  role in global warming, 70
University of California (UC)
  Fruitvale Transit village and, 184–187
  revitalization efforts of, 174–181
Upper Elk River Valley Compact, 201–202
Urban Agricultural Strategy (UAS)
  challenges/obstacles, 164
  design of, 155–159, 162
  industrial ecology and, 163
  place and, 163
Urban agriculture
  civic environmentalism and, 162–165
  DSNI and, 153–154, 155
  historical aspects, 153
Urban life, with natural settings, 49
Urban renewal programs, 31
Urban sprawl, 53–54, 169, 191–192
"Urban villages," 11

Vehicle miles traveled (VMT), 70–71

Walter, E. V., 48
Waste production per capita, 224
Water pollution
  air pollution control and, 59–60
  consumption and, 58–59
  environmental indicators of, 224
  impact, 130
  occupational exposures and, 63
  toxic materials and, 65–68
  transportation and, 56
Watson, Greg, 143, 154–155, 156, 162, 163
Wealth primary system, 35

West, Cornel, 30, 40, 128–129
West Orange, New Jersey
historical aspects, 210
Llewellyn Park, 210–211
Wetlands degradation, toxic materials and, 65–68
Whitman, Christine Todd, 209, 227
Whitman, Walt, 26
Whitt, Laurie Anne, 31
Wilderness Society, 95–96, 110–111
Williams, Jamie, 198, 207
Williams, Raymond, 23
Wilson, William Julius, 37
Wolfe, Alan, 34, 43
Wood, Gordon, 26
Wordsworth, William, 93

Yaro, Robert, 75
Yosemite National Park, Hetch Hetchy Valley, 94

Zeller, Marty, 200, 201, 204